Optimization of Regional Industrial Structures and Applications

Systems Evaluation, Prediction, and Decision-Making Series

Efficiency of Scientific and Technological Activities and Empirical Tests
Hecheng Wu, Nanjing University of Aeronautics and Astronautics
ISBN: 978-1-4200-8846-5

Grey Game Theory and Its Applications in Economic Decision-Making
Zhigeng Fang, Nanjing University of Aeronautics and Astronautics
ISBN: 978-1-4200-8739-0

Hybrid Rough Sets and Applications in Uncertain Decision-Making
Lirong Jian, Sifeng Liu, and Yi Lin, Nanjing University of Aeronautics and Astronautics
ISBN: 978-1-4200-8748-2

Irregularities and Prediction of Major Disasters
Yi Lin, Nanjing University of Aeronautics and Astronautics
ISBN: 978-1-4200-8745-1

Optimization of Regional Industrial Structures and Applications
Yaoguo Dang, Sifeng Liu, and Yuhong Wang, Nanjing University of Aeronautics and Astronautics
ISBN: 978-1-4200-8747-5

Systemic Yoyos: Some Impacts of the Second Dimension
Yi Lin, Nanjing University of Aeronautics and Astronautics
ISBN: 978-1-4200-8820-5

Theory and Approaches of Unascertained Group Decision-Making
Jianjun Zhu, Nanjing University of Aeronautics and Astronautics
ISBN: 978-1-4200-8750-5

Theory of Science and Technology Transfer and Applications
Sifeng Liu, Zhigeng Fang, Hongxing Shi, and Benhai Guo,
Nanjing University of Aeronautics and Astronautics
ISBN: 978-1-4200-8741-3

Optimization of Regional Industrial Structures and Applications

Yaoguo Dang · Sifeng Liu · Yuhong Wang

CRC Press
Taylor & Francis Group
Boca Raton London New York

CRC Press is an imprint of the
Taylor & Francis Group, an **informa** business
AN AUERBACH BOOK

Auerbach Publications
Taylor & Francis Group
6000 Broken Sound Parkway NW, Suite 300
Boca Raton, FL 33487-2742

Printed in the United States of America on acid-free paper
10 9 8 7 6 5 4 3 2 1

International Standard Book Number: 978-1-4200-8747-5 (Hardback)

Library of Congress Cataloging-in-Publication Data

Dang, Yaoguo.
Optimization of regional industrial structures and applications / Yaoguo Dang, Sifeng Liu, and Yuhong Wang.
p. cm. -- (Systems evaluation, prediction, and decision-making series)
Includes bibliographical references and index.
ISBN 978-1-4200-8747-5 (alk. paper)
1. Industrial organization (Economic theory) 2. Economic development. 3. Industrialization 4. Industrial organization--China--Case studies. 5. Economic development--China--Case studies. 6. Industrialization--China--Case studies. I. Liu, Sifeng. II. Wang, Yuhong, 1978- III. Title.

HD2326.D36 2010
338.60951--dc22 2010025851

Visit the Taylor & Francis Web site at
http://www.taylorandfrancis.com

and the Auerbach Web site at
http://www.auerbach-publications.com

Contents

Preface

Industrial structure optimization is a process of coordinating the development of different industries and meeting the increasing demand of the society by better combining and allocating resources. The practice of economic development and industrial transfer has proved that continuously promoting and rationalizing industrial structure has two advantages: accelerating the speed of economic development and realizing more economic benefits.

This book is based on several scientific research projects on optimizing and upgrading industrial structure and many case studies. Briefly speaking, it is the result of the authors' long efforts in the field of industrial structure optimization and its applications. The book contains two parts divided into 12 chapters. The first part (Chapters 1 through 8) is a brief theoretical introduction of regional industrial structure optimization, including the production and development of industrial structure theory, evolution laws of industrial structure, rationalization and upgrading of industrial structure, main factors of industrial structure upgrading, input–output analysis of industrial structure, regional industrial structure and selection of regional leading industries, and the mathematical model of regional industrial structure. The second part (Chapters 9 through 12) includes four case studies. They are approaches and measures of China's industrial structure upgrading during the "11th five-year plan"; the key points, ideas, measures of Jiangsu Province's industrial structural adjustment; main paths and strategies of Jiangsu Province's industrial structural adjustment; and a study on optimization and an upgrading of the industrial structure in M City, respectively.

This book mainly applies grey systems theory to analyze problems on the basis of traditional methods of industrial structure optimization, which combines a new subject—grey systems theory and the theory of industrial structure optimization. We also bring out several models of grey regional industrial structure, including grey correlation priority analysis, an industrial structure order degree measurement model, a regional leading industries grey assessment model and turnpike model—an industrial structure adjustment model. These models not only develop application fields of grey systems theory, but also enrich the theories of industrial structure optimization.

Readers are welcome to point out and correct any errors and omissions in this book.

Yaoguo Dang

Acknowledgments

The relevant researches done in the book were supported by the National Natural Science Foundation of China (No. 90924022 and No. 70901041), the Key Project of Philosophic and Social Sciences of China (No. 08AJY024), the Key Project of the Soft Science Foundation of China (2008GXS5D115), the Foundation for Doctoral Programs (200802870020), the Foundation for Humanities and Social Sciences of the Chinese National Ministry of Education (No. 08JA630039), the Key Project of the Soft Science Foundation of Jiangsu Province (No. BR2008081), and the Foundation for Humanities and Social Sciences of Jiangsu Province (No. 07EYA017). The authors would like to acknowledge the partial support of the Science Foundation for the Excellent and Creative Group in Science and Technology of Nanjing University of Aeronautics and Astronautics and Jiangsu Province (No. Y0553-091). The authors appreciate and thank Professor Yi Lin for his help and Sheng Zhu, Yanyu Wang, Rui Ji, Wenting Wang, and Yu Chen for their assistance.

Authors

Dr. Yaoguo Dang received a bachelor's degree in mathematics from Henan University in 1986, China. In 1991, he received a master's degree in applied mathematics from Zhengzhou University, China and a PhD in management science and engineering from Nanjing University of Aeronautics and Astronautics in 2005.

Dr. Dang is the director of the Department of Management Science and Engineering at the College of Economics and Management, Nanjing University of Aeronautics and Astronautics. He is the vice director of the Institute for Grey Systems Studies and also of the Post-Evaluation Center in Jiangsu Province. He also serves as the vice president of the Grey Systems Society of China. He is the secretary-general of the Chinese Agricultural Systems Engineering Youth Seminar. He is also an advisor for doctoral students in management science and engineering disciplines.

Dr. Dang has directed and participated in more than 30 research projects at provincial and national levels. He has published more than 100 research papers in national and international journals and more than 9 books. Over the years, he has been awarded national and provincial prizes nine times for his outstanding achievements in scientific research and applications. He has won many accolades for his excellent achievements. He was selected as the excellent young scholar of Henan Province in 1998, and was recognized as the best trainer by the "Qing Lan project" of Jiangsu Province in 2004. In 2006, he was selected as the trainer by the "333 high-level talents project" of Jiangsu Province. He was selected as the young leader in science and technology of "333 high-level talents" of Jiangsu Province in 2007.

Dr. Dang serves as a member of the editorial board of some professional journals such as *Journal of Grey System* (Taiwan), *Grey Systems: Theory and Applications* (USA).

Dr. Sifeng Liu received a bachelor's degree in mathematics from Henan University, China in 1981, an MS degree in economics, and a PhD in systems engineering from Huazhong University of Science and Technology, China, in 1986 and 1998, respectively. He was a visiting professor at Slippery Rock University in Slippery Rock, Pennsylvania and Sydney University in Sydney, New South Wales, Australia. At present, Professor Liu is the founding director of the Institute for Grey Systems Studies and the dean of the College of Economics and Management of Nanjing University of Aeronautics and Astronautics. He is also a distinguished professor and guide for doctoral students in the management science and systems engineering disciplines.

Dr. Liu's main research activities are in grey systems theory and regional technical innovation management. He has directed more than 50 projects at the national, provincial, and ministerial levels, has participated in international collaboration projects, and has published over 200 research papers and 16 books. Over the years, he has been awarded 18 provincial and national prizes for his outstanding achievements in scientific research and applications. In 2002, he was recognized by the World Organization of Systems and Cybernetics. Dr. Liu has won several awards such as the "National Excellent Teacher" in 1995, "Excellent Expert of Henan Province" in 1998, "Excellent Science and Technology Staff in Jiangsu Province" in 2002, "Expert Enjoying Government's Special Allowance" in 2000, "National Expert with Prominent Contribution" in 1998, "National Advanced Individual for Returnee" and "Achievement Award for Returnee" in 2003, and "Outstanding Managerial Personnel of China" in 2005.

Dr. Liu is a member of the evaluation committee of the Natural Science Foundation of China (NSFC), a member of the standing committee for teaching in management science and engineering of the Ministry of Education, China. He also serves as an expert on soft science at the Ministry of Science and Technology, China. Professor Liu currently serves as the chair of the Technical Committee of the IEEE SMC on Grey Systems; as the president of Grey Systems Society of China (GSSC); as vice president of Chinese Society for Optimization, Overall Planning and Economic Mathematics (CSOOPEM); as co-chair of the Beijing Chapter and Nanjing Chapter of IEEE SMC; as vice president of the Econometrics and Management Science Society of Jiangsu Province (EMSSJS); and as vice president of the Systems Engineering Society of Jiangsu Province (SESJS); and as a member of the Nanjing Decision Consultancy Committee. He serves as the editor-in-chief of *Grey Systems: Theory and Application*, a member of the editorial board of over 10 professional journals, including *The Journal of Grey System* (UK); *Scientific Inquiry* (USA); *the Journal of Grey Systems* (Taiwan); *Chinese Journal of Management Science*, *Systems Theory and Applications*; *Systems Science and Comprehensive Studies in Agriculture;* and *the Journal of Nanjing University of Aeronautics and Astronautics.*

Dr. Yuhong Wang received a master's degree and PhD in management science from the College of Economics and Management, Nanjing University of Aeronautics and Astronautics in 2006 and 2010, respectively. He is currently with the School of Business, Jiangnan University, China. He was in the Department of Industrial Engineering, University of Arkansas, Fayetteville, Arkansas from August 2008 to August 2009 in a joint training PhD student program between the University of Arkansas and the Nanjing University of Aeronautics and Astronautics sponsored by China Scholarship Council.

Yuhong's research interests focus mainly on grey systems theory, evidential reasoning, and their applications in industrial economics and reliability engineering. He has participated in research projects and published research papers in international journals such as *Expert Systems with Applications*, *the International Journal of Computational Intelligence Systems*, *the Journal of Grey Systems*, *System Engineering Theory and Practice*. He also has participated in international conferences such as *IEEE SMC*, *RAMS*, and *IEEE GSIS*.

1

Origin and Development of Industrial Structure Theory

1.1 Meaning of Industrial Structure

Structure means collocation and arrangement of components from the whole, which was applied in natural science research in the early days. In the field of economics, the phrase "industrial structure" appeared during the 1940s. When it was used to explain the economic problems at that time, the meaning was not exactly clear. It describes the relationship in a certain industry and the relationship among industries. It also means the relationship among enterprises of industries and regional distribution of industries. With the development of research in industrial economy, the meaning and research fields were identified. Industrial structure theory gradually formed a system and the phrase, "industrial structure," was distinguished from industrial organization and industrial distribution.

A normal explanation of industrial structure is the ratio or other relationships of industries in an area or country. There are different ways of understanding industrial structure that have different emphasis points in academic fields. American economist Bain in his book *International Comparison of Industrial Structure* (1966) pointed out that industrial structure was the relationship of enterprises of a certain industry. Then in the book *Industrial Organization* the relationships of enterprises of a certain industry were identified as "industrial organization." In the 1970s, Japanese economists defined the industrial structure as a relationship among industries. Some Chinese scholars hold the idea that industrial structure reflects the relationship of resources distribution, which means the distribution situation of resources (including natural resources and human resources) in different departments of our society; some think that the industrial structure is the distribution of advanced industries; the rest think that it is the structure of national economic departments, which means the structure of all departments and their composition. To research, industrial structure usually means the study of the relationship of production means and livelihood means. Specifically, they are agriculture, light industry, heavy industry, construction industry, and commercial and service industry.

At present, a common explanation of industrial structure is the relational structure of all industries, which means the technical economic connection among industries and the modes of connection. We can investigate this connection and its mode from two angles; one is from the quality angle whereby the tendency of technical economic connections among industries and the mode of connection is revealed dynamically. The investigation shows the substitution law and the structural benefit of leading industries with the development of national economy. The other is from the quantity angle by which the quantity proportion of technical economy of connection and the mode of connection is analyzed. Briefly, it means the ratio of the input and the output in different industries.

Industrial structure research can be divided into broad sense and narrow sense, according to the difference in its connotation and extension. The narrow sense includes the industry types, combination modes which consist of the whole industries, the essential relationship of industries, the technical foundation, development situation of industries, and their roles in the national economy. The broad sense includes the proportion relationship and distribution structure of industries besides the content of the narrow sense.

This book emphasizes the proportion relation of all industries in the national economy when studying the optimization and upgrading of regional industrial structure. Seen from the angle of economics, the proportion relation includes not only the ratio of the input (distribution relation of resources) but also the ratio of the output. The connotation of the industrial structure has complex levels, because both the input and output vary. The input includes resources, labor, capital, and technology; the output can be divided into products and service. Figure 1.1 shows the connotation of industrial structure. The narrow sense is the proportion structure of the industrial structure, and the broad sense covers the input and output structure decided by the

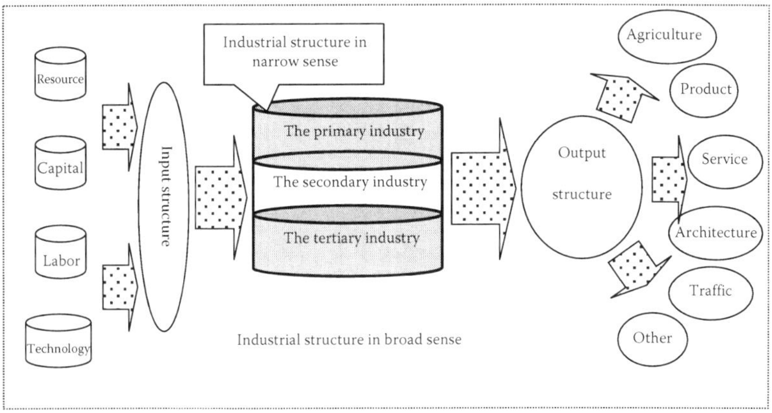

Figure 1.1 Connotation of industrial structure.

industrial structure. Seen from a static angle, the industrial structure is a series of hierarchical proportion relations; seen from a dynamical angle, it is a progress of input and output, which is considered as a kind of resource converter. As a result, we usually use two kinds of indicators to express industrial structure. One is the comparison indicator of input of production factors such as labor and capital. This kind of indicator explains the industrial structure by comparing the resource distribution of the industries. The other kind is the indicator that compares the output of different industries, such as additional values and physical quantities, which show the industrial structure by comparing the operating results of industries.

From the above analysis we find that an important function of the industrial structure is to realize an effective allocation of resources and labor by self-adjusting. We can get maximum benefit by limiting the resources and labor. In the progress of economic development, the status of the industrial structure at different stages are different. In other words, the proportion relationship of industries in the national economy changes from time to time. The change in the industrial structure is the result of economic development, and it also promotes or inhibits economic growth. Therefore, the industrial structure and economic growth are closely related. The development of industrial structure will promote the economic growth and vice versa, which has been proved in the history of many countries. Especially the relation between industrial structure and economic growth is more and more obvious in the modern society. So the adjustment and upgrading of industrial structure are results of economic development of one country or region, and it is also the precondition for economic development. History has proved that the total benefits of different industrial structures are different, which results in different speeds of economic growth. What is more, various situations of economic development lead to different requirements of industrial structure. As a result, all countries and regions of the world are seeking the best industrial structures which benefit them most.

1.2 Origin of Industrial Structure Theory

1.2.1 Ideological Origin of Industrial Structure Theory

The industrial structure theory was established when people studied economic theory and got to the level of industrial structure. Considering the economic problems from structure, the origin of industrial structure theory dates back to the book published in 1672, *Political Arithmetic*, written by William Petty, founder of Britain's classical political economics. Petty first found that the difference of national income and different stages of economic development were mainly affected by different industrial structures. Therefore, he was the first scholar who noticed the structural problems of economy. He said in his book, "Comparing with agriculture, industry has much more income and commerce gets it more than industry." This is called Petty theorem, which shows the direction of structure evolution and economic development. François Qnesnay, representative of French classical political economics and founder of physiocracy, published *Le Tableau économique* in 1758 and *Analyse du tableau économique* in 1766. He analyzed the circulation of total production and reproduction in France. Also, he put forward division of social-stage structures according to his theory of "pure production." He first comprehensively analyzed the origin of every stage, the exchange of capital and income, the exchange of commodity and money, production consumption and living consumption. What is more, Qnesnay considered the circulation of agriculture and industry as a basic factor of reproduction. All these laid the foundation for researches of national economic structure and industrial economic structure. In 1841, German nationalistic economist Georg Friedrich List published the book *The National System of Political Economy*. In this book, he brought forward the "five stage" theory of industrial structure evolution, "all countries have the following development stages: the primitive barbarism period, the animal husbandry period, the agriculture period, the agriculture and industry period, the agriculture industry and commercial period." He also held the idea that state intervention could push the industrial structure optimization. List's theory of industrial evolution and his idea of state intervention had great effect on the development of industrial economics. Some of his opinions were widely used in Japan, Korea, and many other countries. Karl Marx began to write *Das Capital* at the end of 1857 and published the first volume in 1867. Many theories proposed by Marx, such as social reproduction theory, theory of two Category Divisions' balance development, increased priority of means of production were quite special, and they all had direct effect on industrial structure in socialist countries. Léon Walras of The Lausanne School published his book *Elements de economie politique pure*. In this book, the author proposed the theory of marginal utility value. Then he constructed general equilibrium models based on the theory. And in this theory the author applied mathematical models to investigate relations and requirements among industries and sections in national economics from exchange, production, capital formation, currency circulation aspects. This theory is also one of the theoretical fundamentals for input and output analysis.

1.2.2 Formation of the Industrial Structure Theory

Modern industrial structure theory was formed in the 1930s and 1940s. During this period, many economists made outstanding contributions to the formation of this theory, including the German economist Hoffman, Japanese economist Akamatsu, British economist Clark, American economists Kuznets and Leontief.

German economist Hoffman published the book *The Stage and Type of Industrialization* in 1931, in which he investigated evolution laws of the structure of industrial sector during the industrialization, and proposed the "Hoffman theorem." In his view, during the process of industrialization, "Hoffman coefficient" (namely the ratio of net production value between consumer materials and capital materials) decreases; thus, he revealed the heavy-industrialization trend in the industrial structure.

In 1935, Akamatsu put forward the "flying geese patterns theory" about industrial development. The theory argued that the industrial development of a country should work closely with the international market to make the internationalization of the industrial structure. In his view, the countries developing late can speed up their process of industrialization by four stages; the industrial development policies should be formulated according to characteristics of flying geese patterns.

British economist Clark published the book *The Conditions for Economic Progress* in 1940 in which he summed up, through data collation and comparison of total labor input and output of more than 40 countries and regions in three major industries in different periods, that there is a certain degree of regularity between the structural changes of labor force in the three major industries and the improvement in per capita national income. In his view, along with the economic development (that is, the improvement of per capita national income), the employment proportion of the primary industry will continue to decrease, while those of the secondary and tertiary industries will increase, which reveals that the labor force transfer between different industries results from the relative income differences in each industry in the process of economic growth. This result is known as "Petty–Clark theorem" which created a new direction in the industrial structure research.

Leontief was a well-known American economist, who was the founder of input–output analysis method and received the Nobel Prize in economics in 1973. His research on the industrial structure has a considerable influence. In 1941, he made an in-depth and systematic analysis on the economic structure of the United States. His book *Economic Structure in the United States from 1919 to 1929* is one of the classic works in the industrial structure theory.

From this stage onward, economists and scholars started to change their research methods, from initial empirical analysis to theoretical research, promoting the formation of the industrial structure.

1.3 Development of the Industrial Structure Theory

The industrial structure theory rapidly developed in 1950s and 1960s. During this period, there were many representatives who made great contributions to the industrial structure theory including Leontief, Kuznets, Lewis, Hirschman, Rostow, Chenery, Hoffman, Tinbergen, Higgins, and so on.

Leontief carried out a more in-depth research on the industrial structure based on the initial study. He published two books, *Research on the U.S. Economic Structure* in 1953 and *Input–Output Economics* in 1966 and set up a system of input–output analysis, including input–output analysis method, input–output model, and input–output table. He applied this kind of method to analyze the relationship between the structure of an economic system and each department in the production and investigated the impact of dynamic development and technological changes on the economy; he also analyzed the relation of foreign trade and domestic economy and finally the economic relations across regions and impact of economic policies in the country. The input–output analysis today has developed greatly since its conception. In-depth, developments are shown by endogenous change of exogenous variables, static models developing to dynamic models, and the optimization model made through a combination of input–output analysis, linear programming, nonlinear programming, and dynamic programming. In breadth, the model is expanded to environmental pollution control model, international trade model, the world model, population and education model, and so on. The basic idea is that to obtain a certain output, there must be an input. It is a programming approach to balance products across departments comprehensively. And it is also a kind of economic analytical method to analyze economic issues such as economic structure, economic efficiency, economic policy, and product price comprehensively, which can be used for analysis of the national economic structure. Now there are more than 100 countries and regions using this method to work out their input–output table. The input–output analysis method is the most common mathematical tool which is used to analyze industrial structure all over the world.

Kuznets carried out a more in-depth research on the relationship between the economic growth and industrial structure. Based on Clark's research results, he gathered statistics of dozens of countries. From the aspects of distribution of national income and labor force in industries, his study on industrial structure had gone deep into the ratio and change of national income brought by three industries. Through empirical analysis on changes in industrial structure, Kuznets concluded that the industrial structure may have new changes after developed countries had entered the stage of modern economic growth. As the economy develops, the national income or GNP of the primary industry will take up lesser proportion of the national income; so does the labor force of the primary industry, which means that agriculture plays a decreasing role in the national economic growth. The proportion of the national income or GNP of the secondary industry will have a slight increase along with the economic development, while that of the labor force will maintain the same or have a slight increase, which means that the industry will contribute more to economic growth. With the economic development, the national income or GNP of the tertiary industry has a slight increase, but not a consistent one, while the labor force continues to rise. As can be found from two important representative books of Kuznets, *Modern Economic Growth* (1966) and *Economic Growth of Countries* (1971), in the analytical framework of modern economic growth, the emphasis is the study on the general trend of structural changes from which he arrived at the above important conclusions.

Lewis, one of the founders of American development economics and a Nobel laureate in economics of 1979, published a well-known thesis "Economic development under the unlimited supply of labor" in *Manchester Journal*, putting forward the dual economic structure model, that is, the whole economy is composed of a small modern capitalism department and a strong traditional agriculture department. The model is used to explain the economic problems of developing countries. Traditional agriculture department is mainly self-sufficient in agriculture, retail trade, and most of the traditional sectors, which is characterized by a low marginal labor productivity and a large number of surplus labor force. With the exception of modern industry, modern capitalist department includes a small number of high-efficiency agriculture and modern business, where the labor productivity is relatively high and the employment small, the labor required being gradually transferred from the traditional agricultural sector. Economic development need to expand the modern capitalist department, narrow the traditional agricultural department, obtain unlimited cheap labor supply from the traditional sectors, and gain huge profits from the difference between the price of labor supply and marginal labor productivity. Developing countries can make use of the favorable conditions of rich labor to speed up the economic development. In 1958, Lewis published the book *Theory of Economic Growth*, in which he comprehensively analyzed the economic and noneconomic factors that affect the economic development, including the capital accumulation, technological progress, population growth, social structure, economic system, religion, cultural and historical traditions, polity and psychology, and so on. He mainly discoursed upon economic structure, especially upon the industrial structure, with rich content and profound exposition.

Hirschman, the American development economist and the founder of the famous "unbalanced" growth theory, published one of the most influential books on economic development issues in 1958, *The Economic Development Strategy*. In this book, he rejected the popular opinion prevailing at that time that developing countries must develop according to a carefully controlled balanced growth route and designed an unbalanced growth model. After studying the development experiences of a large number of Latin American countries, Hirschman held the opinion that comprehensive investment and development in all social sectors are almost impossible due to scarcity of resources in developing countries. So they can only selectively put the limited resources into certain sectors to play their role of promoting economic growth farthest. There are two approaches to realize the unbalanced growth: one is the "shortage development," that is, firstly invest in the direct production which will cause a shortage of social capital, and then cause an increase in direct production costs, thereby the investment will be forced to transfer to social capital so that the balance can be obtained. Through investment in direct production, a new round of process of unbalanced growth will be triggered. The other is "excess development," that is, first invest in social capital in order to reduce the cost of direct production activities, thus investing more capital to direct production. When the balance is reached, repeat the process. Economists in the early development limited their theories to development priorities of the direct production sectors and infrastructure sectors. Hirschman broke through those narrow arguements and replaced them with a broad discussion as to whether plans for the national economy should develop some sectors with priority and emphasis. Among them, the theory of "correlation effects" (including the forward and the backward correlation effects) and the theory of "most effective order" have become important analytical tools for economic development.

American economist Rostow put forward two well-known theories: the theory of leading industry spillover effects and the theory of economic growth stage. According to relevant technical standards, he divided economic growth process into six stages, which are traditional society stage, the stage of creating the prerequisite for the take-off, the take-off stage, mature stage, people's high consumption stage and the stage of pursuing quality of life. The evolution of each stage is characterized by replacement of leading industrial sectors. Rostov believes that changes in industrial structure have significant impact on economic growth; in the process of economic development, attention should be paid to spillover effects of the leading industry. Rostov has made norms to spillover effects of the leading departments in the industry. He thinks this effect works as "disproportionate growth" of certain sectors in the growth of various historical stages. Specifically speaking, the spillover effects include: first, recall effects, which mean the impact of growth in the leading departments on departments that provide means of production for them; second, forward-looking effects, which mean the inducement that the leading departments bring to the emergence of new industries, new technology, new quality, and new energy resources; third, side effects, which mean the function of the growing leading departments on the socioeconomic development of the surrounding area. At the same time, Rostov also believes that the socioeconomic development is ultimately motivated by people's subjective tendencies: tendency of the development science, tendency of applying science to economic purposes, tendency of innovation, tendency of seeking material progress, consumption tendency, and childbearing tendency. These theories have a broad influence. His major works like *The Process of Economic Growth* and *Stages of Economic Growth* reflect the outstanding contribution he made to the study of these aspects.

Chenery, a well-known American economist, also made an in-depth research on the problem of the industrial structure. He put forward a "development type" theory, making a great contribution to the development of the industrial structure theory. On the one hand, he further expanded his research areas to low-income developing countries, which took industrialization as its main line since World

War II, especially the development experience of newly industrialized countries. On the other hand, his research idea in empirical studies was also quite different from that of Kuznets. His basic idea was to reveal "standard type" of economic development and structural changes based on in-depth and comprehensive analysis on structural changes and factors that can affect structural changes through various forms of comparative studies. That is, sum up some basic development models and different characteristics of economic development and structural changes of countries, thus drawing some typical truths and conclusions with regularity. Through calculation on 53 different types of national economic statistics in 1960, he obtained a set of standard values about the comparative weight changes in each manufacturing sector along with changes in the level of per capita income, which can be used to comparatively analyze the internal structure of the manufacturing sector under certain economic conditions, that is, whether the industrial structure has deviated from the normal value. The book *Patterns of Development: 1950–1970*, written by Syrquin and Chenery in 1975, was based on 20,000 observation data of 130 variables in the statistics from 1950 to 1970. They analyzed and compared the whole process of economic structure changes commonly experienced by 101 countries in their economic growth processes to form "the patterns of development," which reflect the typical relationships among the main variables of structural changes. Chenery thought that the substitution elasticity of capital and labor did not change in the economic development, thus presenting the production function theory of Cobb–Douglas. He pointed out that the industrial structure would change in the course of economic development, the export of primary products in the foreign trade would decrease, and the import substitution and export substitution would be implemented step by step. The fruits of his research include *Industrial Association Economics*, *Study on Development Plans*, *Patterns of Development*, *Structural Change and Development Policy*, *Comparative Study on Industrialization and Economic Growth*, and so on.

German economist Hoffman studied the evolution of industrial structure during the industrialization process. He introduced structure variable into demand formula, and successfully demonstrated its actual significance in the area of statistics. He also studied the contribution of import substitution to economic development, and made a great contribution to the theories of industrial structure. He put forward the famous "Hoffman theorem"—the ratio between industrial net output of consumption material and capital material during the process of industrialization declines constantly. The ratio is referred to Hoffman ratio. The industry can be divided into four stages according to Hoffman ratio. At the first stage of industrialization, the production of consumer goods is dominated in the manufacturing sector, while capital goods production in the manufacturing industry is not developed; in the second phase of industrialization, capital goods industry grows faster than consumer goods industry, but the scale of the production of consumer goods is still larger than that of capital goods industry; at the third stage of industrialization, the production of capital goods industry continues to grow, expands rapidly, and stays in a state of equilibrium with consumer goods industry; in the fourth stage of industrialization, the production of capital goods is dominant, the scale of which is larger than that of production of consumer goods, basically achieving industrialization. Hoffman ratio and Hoffman industrialization rules are in line with the law of industrial development, especially in the early stage of development of industrialization, showing great value and broad influence.

One of the founders of econometrics is the Dutch economist Tinbergen. His theory on economic policy-making has rich industrial structure theories. For example, in his view, economic policies need a conscious use of some means to achieve certain goals, which contain a way of restructuring. He put economic policies into three types: number policy, nature policy, and reform policy. Among them, the number policy was to use the dynamic model to measure and analyze the evolution of industrial structure; the nature policy was to change some elements of the structure (input–output tables); the reform policy was to change some basic elements. In another example, he adopted large-scale simultaneous equation system in his theory of the development plan, which was a mixture of many models made by Cairns, Harold, Leontief, and so on. In addition, the part of the input–output law he used was an industry association method, which started from investment planning projects directly, and simply added the microplan into a macroplan.

Canadian economist Higgins has made some research on the dual economic structure. He described the characteristics of dual-economic structure of developing countries. In his view, the production function of advanced departments and original departments were completely different. The production function of original departments belonged to an alternative type; because of absolute lack of funds, these departments chose labor-intensive technology; the production function of advanced departments belonged to a fixed input coefficient type, and these departments adopted a capital-intensive technology. The major works reflecting his research results include *Indonesia's Economic Stability and Development*, *Economic Development: Principles, Problems and Policies*.

Shinohara Miyohei was a well-known expert studying economic cycle theory and industrial structure theory in Japan. In mid-1950 he proposed two benchmarks for planning industrial structure of Japan income-elasticity benchmark and productivity-increase benchmark. The requirement of income-elasticity benchmark was to put the accumulation into the industries or departments, which had large income elasticity. Because these industries or departments had a larger market demand, it was convenient to take advantage of large scale economic benefits to improve profits quickly; productivity-increase benchmark required the accumulation to be put into the industries and departments whose productivity increased the fastest, as the productivity of these industries or departments increased relatively quickly and their unit cost also had a quicker decline; at a certain wage rate, these industries and departments had the largest rising profit margin rate. As basic standards of industrial structure planning, the two benchmarks were quite concise and effective in identifying and grasping basic characteristics of leading industries. The benchmarks were used by Japanese industrial structure council for constituting industrial structure policies in 1963 and medium-term economic plans in 1965. Correspondingly, Japanese economists also suggested two concepts: influence coefficient and sensitivity coefficient of industries, which were used for the analysis on correlation between industries and accepted by economists of all countries. In addition, Japan had to establish domestic social reproduction cycle system on the basis of international trade to achieve industrial structure sophistication. Therefore, Shinohara Miyohei proposed the famous "dynamic comparative cost theory." In his view, from a development perspective and dynamic point of view, although some products were disadvantaged in international trades at a certain point of time, it is possible for them to be transformed into advantaged products after a certain period of time. Especially, the comparative cost of products can be transformed. As for industries with development potential and having significant impact on the national economy, they could become strong export industries and be able to obtain the dynamic comparative advantage after 10 years of support. This is the so-called supporting small industries opinion, and is seen as a theory of "win by striking only after the enemy had struck," which makes a significant reference to industrial development of late starting countries.

1.4 Research Field and Theoretical System of Industrial Structure

1.4.1 Research Field of Industrial Structure

Industrial structure theory regards technological and economic relation and its contact mode among industries as its research object. It mainly studies industrial constitution structure with links among industries and relates between each industry and the overall economy within a certain economic system. Through studying the history, current situation, and future research of industrial structure, the general trend of industrial structure development and changes are revealed. The basic modalities of relation structure among industries are mainly as follows:

1. *Industrial structure in economic growth.* Industrial structure in economic growth means an association relationship between industries formed in the process of economic growth of a country or a region during a certain period and the dynamic changes of the association relationship. In the process of socioeconomic growth, it requires a coordinated development between industries in the social reproduction process. The evolution of industrial structure and economic growth are closely linked intrinsically.
2. *Output structure.* Output structure refers to the ratio and distribution of each industry's output in the total input of the total national economic output. Social demand structure, trade structure, and productivity development level determine the social output structure; the relation of each industry's output in the total national economic output reflects the

relative status of this industry in the national economy, and its dynamic changing process reflects this industry's growth from another aspect.

3. *Input structure.* Input structure refers to the ratio and distribution of each industry's input in the total input of the total national economic output. Input structure of an industry can reflect the industry's demand structure from one side, and can be further decomposed into consumer demand and investment demand. Consumer demand covers an industry's demand for life and production; investment demand includes an industry's need for capital, labor, technology, and other factors of production. The research on the industrial input structure not only analyzes the quantitative structure relationship, but also studies how quality changes of input factors lead to quantitative structure changes. The investment structure research of industries is one of the most fundamental problems of industrial structure research, which includes the stock structure and increment structure. And the adjustment of the investment structure is the basic means of optimizing industrial structure.
4. *Input–output structure.* Input–output structure means the input–output state of each industry in the national economic input–output balance, which is based on the research of input and output structure among industries. Input–output structure reflects the efficiency and effectiveness of the operation of the economic system. And the industrial structure is the input–output converter. The research on input industrial structure among industries is designed to reveal the inner converting mechanism of industrial structure.
5. *Regional distribution structure of industries.* The regional distribution structure of industries refers to the spatial distribution of each industry in a certain economic region. And it is affected by each industry's input–output characteristics, regional resources and their consumption characteristics, inter-regional trade exchange conditions, industrial policy, and other factors. Economic region system is an open system; a small economic region is a subsystem of a greater economic region, which participates in the economic division of large-scale systems, and also has certain independence. Industrial

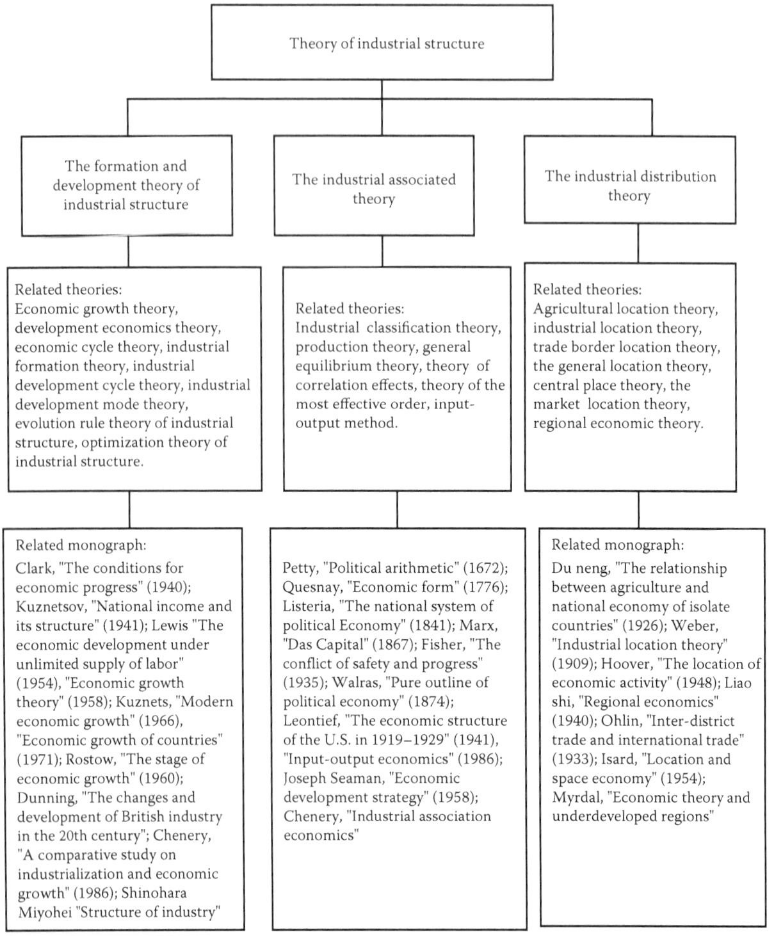

Figure 1.2 Theoretical framework and articles of industrial structure.

structure within the region affects the region's economic status and economic growth.

1.4.2 Theoretical System of the Industrial Structure

The method of allocating resources effectively and rationally to achieve a sustained and steady economic growth is one of the major issues of the economics study. And it is also the main clue of industrial structure theoretical research. The fundamental system of industrial structure includes formation and development theory of industries, industrial associated theory, and industrial distribution theory.

Formation and development theory of industries includes theory of industrial structure formation, theory of industrial structure evolution, and theory of industrial structure optimization, and so on. The theory of industrial structure formation is the study of some basic concepts and research areas of industrial structure as well as the formation conditions, mechanism and process. The theory of industrial structure evolution is study of theoretical problems like internal relations of evolution of industrial structure, economic growth and economic development, status and trends of industrial structure, general trend of industrial structure evolution, factors affecting industrial structure, evolution cause of industrial structure, and empirical study on industrial structure evolution in different countries. The theory of industrial structure optimization is the study of industrial restructuring and its function mechanism, industrial structure upgrading and its function mechanism, industrial structure rationalization and its function mechanism, selection of leading industries, and selection and conversion of the strategic industries.

The industrial associated theory is the study of the associated mode of industries by methods such as input–output analysis to investigate associated quantitative relations between industries, effects of industries, and other theoretical issues.

The industrial distribution theory is the study of theoretical issues such as conditions, characteristics, levels of industrial distribution, and regional industrial structure. Relative theories and work about industrial structure are shown in Figure 1.2.

2

Evolution and Laws of Industrial Structure

2.1 Related Theories on Industrial Structure Evolution

2.1.1 Conceptions of Three Industries

Fischer, a British economist and a professor at the University of Otago in New Zealand put forward conceptions of *Three Industries* in his book *Security and Progress* in 1931. He analyzed the history of economic development as the emergence, development, and evolution of three industries and highlighted the evolution of industrial structure in the economic development. In his view, the history of economic development can be divided into three stages. In the first stage, the production activities of the humans were based on industry and livestock; in the second stage, they were assigned for the development of large-scale industries; in the third stage, with large amount of capital and labor flowing in, they were directed toward nonmaterial production. In order to adapt to the three major stages of human economic development there was the division into three industries. The primary industry was agriculture and animal husbandry, which was characterized by the direct use of natural resources to provide basic means of subsistence—food; the secondary industry was the manufacturing industry, whose basic characteristics were reprocessing of natural resources, and primarily provided material products other than food; the tertiary industry was the service industry, which was characterized by not providing tangible goods, but mainly meeting the needs of a wide range of service industries beyond material needs.

Colin Clark widely made use of the three industries concept in the book *Conditions for Economic Progress*, which was used in analyzing industrial structure in order to understand the changing characteristics and laws of industrial structure in the process of economic development from the divisions into three industries and their relationships.

Thus, the idea of the three industries and division method spread. And the three industries classification method came to be known as the Great Clark classification.

Fisher, Clark, Simon, Kuznets, Chenery, and others used three industries classification method to analyze the process of economic development. They studied the evolution and the law of industrial structure from the view of changing industrial structure given the stage of economic development and its changing law. In particular, the Kuznets' positive analysis about evolution of structure on the history of multinational economic development as well as the standard model analysis on multinational structure evolution of Chenery and others—which made the relationship between economic development and structural changes more comprehensible—clearly reflected the requirements of economic development to the different historical stages of the structural changes. Thus, they contributed in providing depth to the theory of industrial structure.

2.1.2 Law of Three Industrial Structure Changes: Petty–Clark Theorem

In the 17th century, the British economist, William Petty in his famous work *Political Arithmetic* described this phenomenon: the manufacturing industry could get more income than agriculture industry; what is more, the commerce industry could get more income than manufacturing industry. At that time, most of the population in The Netherlands were engaged in manufacturing and business; the per capita income was much higher than that in other European countries. Such disparity in income between different industries resulted in the labor force transferring to higher income departments. Here, Petty had just described the phenomenon of labor force transferring between industries and pointed out that labor tended to shift to the higher income departments; because of age restrictions, the author in the theory could not accurately demonstrate the relationship between per capita GDP growth and industrial conversion.

In the 1950s, the British economist Clark empirically summarized the relationship between the per capita GDP growth and industrial conversion for the first time. In his book, *The Conditions for Economic Progress*, he collected and compiled the total output data and time-series data on labor into different departments from more than two dozens of countries and revealed, by statistical analysis method, the general laws of industrial structure changes in the process of the economic progress. With per capita income level rising, labor forces first transferred from the primary industry to the secondary industry; when the per capita national income increased further, the labor forces then transfered to the tertiary industry. Clark thought that the law he had found was just the proof of Petty's point of view, so he called it Petty theorem. Later, Clark's discovery was known as the Petty–Clark theorem.

There were three preconditions in Clark's analysis. First, put all economic activities into the basic framework of the primary industry, the secondary industry, and the tertiary industry by using the three industrial classification method. Second, analyze the distribution of labor force in the different industries when studying the evolution of industrial structure with economic development. Third, explore the law of the evolution of industrial structure based on the per capita income time-series data in a number of countries.

Petty–Clark theorem's empirical conclusion could be confirmed not only from the time-series analysis about the economic development of a country, but also from a cross-sectional comparison in a country at different levels of development at the same point, as shown in Table 2.1. Namely, the higher the level of per capita income, the smaller the agricultural labor proportion in the total labor force. However, the proportion in the secondary and tertiary industries would be relatively greater than that in the primary industry and vice versa.

2.1.3 Kuznets Laws

The Petty–Clark theorem mainly points out the evolution law of labor force distribution structure in the primary, secondary, and tertiary industry with economic development and the reason for the labor force distribution structure change was due to the relative income difference produced during the economic development. Similarly, research on the law of industrial structure evolution must be extended to the variation of the national income and its distribution structure in the primary, secondary, and tertiary industry. If trends of the national income distribution are known in three industries, we could combine together distribution trends of labor

Table 2.1 Levels of Economic Development and Industrial Structure of Countries (1960–1990)

YEAR	COUNTRIES (REGIONS) GROUP OF DIFFERENT LEVELS OF ECONOMIC DEVELOPMENT	PROPORTION OF DIFFERENT INDUSTRIES IN OUTPUT VALUE (%)			EMPLOYMENT PROPORTION OF DIFFERENT INDUSTRIES (%)		
		1	2	3	1	2	3
1960	Low-income countries	48	25	27	77	9	14
	Under middle-income countries	36	29	39	71	11	18
	Middle-income countries	24	30	46	62	15	23
	Above middle-income countries	18	33	49	49	20	31
	Developed countries or regions	6	40	54	18	38	44
1981	Low-income countries	37	34	29	73	13	15
	Under middle-income countries	22	35	43	54	17	29
	Middle-income countries	14	38	48	44	22	34
	Above middle-income countries	10	39	51	30	28	42
	Developed countries or regions	3	36	61	6	38	56
1990	Low-income countries	31	36	35	–	–	–
	Under middle-income countries	17	31	50	–	–	–
	Middle-income countries	12	37	50	–	–	–
	Above middle-income countries	9	40	5	–	–	–
	Developed countries or regions	–	–	–	–	–	–

Source: Liu Wei, *Industrial Research in the Process of Industrialization*, p. 29.

force and national income and deepen the analysis of causes of the evolution of industrial structure.

In this aspect, prominent achievements in the United States were made by the well-known economist Kuznets, who was called the "GNP father" in Western countries and was good at the statistics of the national economy. Kuznets wrote the above research results of the theory of industrial structure in his main works, *Modern Economic Growth* and *Economic Growth in Quantity*. Based on Clark's research results, Kuznets collected and presented a large amount of data from over 20 countries on national income and labor distribution and made an analysis of industrial structure change along with the economic development. The trend of the laws of evolution of national income and labor distribution structure among industries has been summarized from a statistical angle as follows:

- The proportion of the national income from the agricultural sector (namely, the primary industry) in the entire national income would drop constantly with time, in the same proportion as agricultural labor in the total labor force.

- The relative proportion in the national income from the industrial sector (namely, the secondary industry) would rise generally; however, the proportion of the labor force in the industrial sector would remain generally unchanged or go up slightly according to the conditions in the different countries.
- The relative proportion of the labor force in the service sector (namely, the tertiary industry) would rise in almost all countries; however, the relative proportion of the national income might not increase in a synchronized fashion with the relative proportion of labor.

2.1.4 Chenery and Others across National Models

As mentioned earlier, the positive studies by Fisher, Clark, Kuznets, and others on the history of economic development empirically summed up three characteristics of industrial structure for different levels of economic development from the perspective of national income and labor force. Based on empirical researches by Kuznets and others, Chenery and others combined analysis of different countries from the point of view of supply, demand, trade, investment, and other points, analyzed more systematically and comprehensively the development characteristics at different stages.

Chenery and others divided the whole process of change, which was from the level of developed economies to the mature industrial economies, into six phases according to per capita national income. In the first phase, the per capita national income is US$ 140–280; in the second phase US$ 280–560; in the third phase US$ 560–1120; in the fourth phase US$ 1120–2100; in the fifth phase US$ 2100–3360; and in the sixth phase US$ 3360–5040. Then he selected representative countries as samples and used multiple regression analysis to solve the Wassily W. Leontief's input–output model. Then he got the structure situation of all time points and built up system expressions about the conversion process of the economic structure at different stages of economic development and characteristics. He established the so-called average or standard industry model. In the simulation analysis of Chenery and others, the economic development (also called the economic conversion process) from a developing stage to a developed stage included only the first four phases of the six phases. That is, the development process of per capita national income from US$ 140 to 2100. The corresponding models are shown in Tables 2.2 through 2.4.

From Tables 2.2 through 2.4, when the per capita national income changes from $140 to 2100 we see that the prominent changes in the supply structure were as follows: the share of per capita national income provided by agricultural sector had dropped from 37% to 6% in the whole per capita national income, a decrease of 31%; however, in the same period, the share of the per capita national income provided by the manufacturing industry increased from 15% to 36% in the entire composition of national income, an increase of 21%. These corresponding changes in the structure represented that the major provider of per capita income of the economic structure had changed from the agricultural sector to the industrial manufacturing sector; economic development had made a fundamental structure quality conversion.

The prominent changes in the structure of demand were as follows: the middle demand took great changes regardless of the demand for final demand or the entire total output. In comparison with the final demand, the middle demand increased from 5% to 82%; compared with the total output, the middle demand increased from 33% to 45%. The great increase of the middle demand shares reflected the high level of the evolution of the industrial structure, improvement of specialization, and improvement of the complexity of the production contact.

The prominent changes in the trade structure were as follows: the proportion of agriculture exports in the total exports dropped from 75% to 20.17%; industrial products exports in total exports rose from 7.14% to 52.19%, of which machinery exports rose from 0% to 10.34%; agricultural products imports in the total imports rose from 9.68% to 17.06%; industrial goods imports in the total imports decreased from 70.98% to 52.87%, in which the machinery imports declined from 25.81% to 22.30%. The meaning of the structural changes is that economic development, which is the core of industrialization, is the process of changing industrial structures of developing countries; at the same time, the skill level of production was increased and its comparative advantage had been changed. So the terms of trade of developing countries had favorably changed.

Table 2.2 Standard Model of Multinational Solutions—Benchmark Level of Income 1 (US$ 140 per capita)

	DOMESTIC FINAL DEMAND (US$)					TRADE (US$)						
DEPARTMENT	CON-SUMER	INVEST-MENT	GOVERN-MENT	AGGREGATE DEMAND	AGGREGATE DEMAND (%)	EXPORT	IMPORT	NET TRADE	MIDDLE DEMAND, *W*	TOTAL OUTPUT, *X*	VALUE ADDITION, *V*	VALUE ADDITION (%)
Primary products												
1. Agriculture	25.5	0	0.5	26	18	21	3	18	17	61	51	37
2. Mining	0.5	0	0.5	1	0	1	1	0	2	3	2	1
Subtotal	26	0	1.0	27	18	21	4	18	19	64	53	38
Manufacturing sector												
3. Food	15	0	1	16	11	0	3	–3	10	23	7	5
4. Consumer goods	11	0	0	11	8	1	4	–3	9	17	8	6
5. Products	3	0	0	3	2	1	7	–6	11	8	4	3
6. Machinery	0	7	1	8	6	0	8	–8	1	1	1	1
Subtotal	29	7	2	38	27	2	22	–20	31	49	20	15
Nontrading sector												
7. Social infrastructure	7	12	1	20	14	1	1	0	7	27	15	11
8. Services	41	1	16	58	40	3	4	–1	13	70	51	36
Total	102	21	20	143	100	28	31	–3	70	210	140	100
Total final demand (%)	71	15	14	100								

Source: Liu Wei, *Industrial Research in the Process of Industrialization*, p. 58.

Table 2.3 Standard Model of Multinational Solutions—Benchmark Level of Income 3 (US$ 560s per capita)

	DOMESTIC FINAL DEMAND (US$)				TRADE (US$)							
DEPARTMENT	CON-SUMER	INVEST-MENT	GOVERN-MENT	AGGREGATE DEMAND	AGGREGATE DEMAND (%)	EXPORT	IMPORT	NET TRADE	MIDDLE DEMAND, *W*	TOTAL OUTPUT, *X*	VALUE ADDITION, *V*	VALUE ADDITION (%)
Primary products												
1. Agriculture	45	0	1	46	8	62	21	41	62	149	101	18
2. Mining	1	0	1	2	0	3	6	−3	19	18	5	3
Subtotal	46	0	2	48	8	65	27	38	81	167	116	21
Manufacturing sector												
3. Food	61	0	4	65	11	6	11	−5	48	108	38	5
4. Consumer goods	51	0	3	54	10	14	13	1	65	120	56	10
5. Products	18	0	1	19	3	15	31	−16	78	81	38	7
6. Machinery	2	36	1	42	8	1	36	−35	13	20	10	2
Subtotal	132	39	9	180	32	36	91	−55	204	329	137	24
Nontrading sector												
7. Social infrastructure	32	65	7	104	18	10	7	3	42	149	84	15
8. Services	161	6	69	235	42	24	17	7	65	307	223	40
Total	371	110	86	567	100	135	142	−7	392	952	560	100
Total final demand (%)	65	20	15	100								

Source: Liu Wei, *Industrial Research in the Process of Industrialization*, p. 59.

Table 2.4 Standard Model of Multinaticnal Solutions—Benchmark Level of Income 5 (US$ 2100 per capita)

	DOMESTIC FINAL DEMAND (US$)					TRADE (US$)						
DEPARTMENT	CON-SUMER	INVEST-MENT	GOVERN-MENT	AGGREGATE DEMAND	AGGREGATE DEMAND (%)	EXPORT	IMPORT	NET TRADE	MIDDLE DEMAND, *W*	TOTAL OUTPUT, *X*	VALUE ADDITION, *V*	VALUE ADDITION (%)
Primary products												
1. Agriculture	64	0	4	68	3	117	101	16	89	253	137	6
2. Mining	19	0	4	23	1	12	78	–66	125	82	62	3
Subtotal	83	0	8	91	4	129	179	–50	294	335	199	9
Manufacturing sector												
3. Food	186	0	19	205	10	59	32	27	180	412	126	6
4. Consumer goods	202	0	15	217	10	91	45	46	332	595	270	13
5. Products	104	0	6	110	5	95	104	–9	480	581	263	12
6. Machinery	15	172	4	191	9	60	132	–72	73	192	99	5
Subtotal	507	172	44	723	34	305	313	–8	1065	1780	758	36
Nontrading sector												
7. Social infrastructure	115	283	26	424	20	41	28	13	150	587	330	16
8. Services	549	25	300	874	42	105	72	33	219	1126	813	39
Total	1254	480	378	2112	100	580	592	–12	1728	3828	2100	100
Total final demand (%)	56	23	18	100								

Source: Liu Wei, *Industrial Research in the Process of Industrialization*, p. 60.

In addition, in the whole structure changes mentioned above, the proportion of the service sector essentially remained unchanged in total demand and in total output. This showed that prior to the completion of industrialization, the structural changes were mainly seen in the relationship between the primary and the secondary industry. This was also established in Kuznets statistics.

2.2 General Trend in the Evolution of Industrial Structure

Industrial structure constantly changes according to the economic development. This change mainly shows the upgrading of industrial structure, evolving from low-level to high-level, and rationalization of the horizontal evolution of industrial structure. These upgrading and rationalization of industrial structure promote economic development. From the practices of many developed countries and newly industrialized countries, there are several laws on the evolution of industrial structure (upgrading and rationalization of the industrial structure) from different points of view.

2.2.1 Stage of Industrialization Development

The evolution of industrial structure has the following stages: preindustrialization, early industrialization, mid-industrialization, industrialization in the late stage, and postindustrialization. In the preindustrialization stage, the primary industry is dominant, secondary industry has certain development, and the tertiary industry's position is weak. In the initial stage of industrialization, the output value of the primary industry in the national economy is gradually narrowing, the status declining; the secondary industry develops further, the industry center gradually shifts from being light-industry-oriented to basic-industry-oriented and the secondary industry is dominant; the tertiary industry has a certain presence, but the proportion is still relatively small in the national economy. In the mid-industrialization, the industry center changes from basic industries to high-degree processing industries, the secondary industry is still in the dominant position, and the tertiary industry gradually increases. In late industrialization, the proportion of secondary industry continually declines, the tertiary industry has a rapid development in which the information industry grows quickly, the proportion of tertiary industry output values is in the dominant position in three industries, and almost in absolute dominance. During the stage of postindustrialization, the knowledge-based industries has become the main industries. The industrial structure develops with such a development process from a low-level to a high degree of modernization.

2.2.2 Conversion Process of Leading Industries

The evolution of industrial structure has several phases such as the agriculture-oriented phase, textile industry-oriented phase, the heavy industry-oriented phase, which have the basic industries like raw materials industries and fuel power industry, and so on as the central industries, low-degree processing-oriented industries phase, high degree processing and assembly-type-oriented industries phase, and the tertiary industry-oriented phase, information industries-oriented phase, and so on. At every stage of the evolution of the industrial structure, the general rules are as follows:

1. In the agriculture-oriented phase, the proportion of agriculture is absolutely dominant, and the secondary industry and the tertiary industry are very limited.
2. In the stage where the textile industry is dominating due to demand and the simple requirements of technology, cheap labor forces are separated from the primary industry, and the light textile industry has a rapid development; the pace of the primary industry development declines, the status weakened; the heavy industry and the tertiary industry develop slowly. At this time, the textile industry replaces the primary industry to become a leading industry.
3. In the heavy industry phase, the basic industries such as raw materials industry and fuel power industry, and so on play a central role, the proportion of output values of agriculture is very small in the national economy; the textile industry continues to develop, however, the pace gradually slows down; the basic industries such as raw materials, fuel power and infrastructure industry, and so on develop rapidly and gradually

replace the textile industry to become leading industries. These basic industries are the forerunner industries or the constrained industries of heavy industries. They must be developed first and then would not become a bottleneck in the development of heavy industries.

4. In the heavy industry-oriented phase of low processing and assembly-type industry that belong to a phase of simple technical requirements, steel, shipbuilding and other low-level processing, and assembly-type heavy industries develop rapidly, and they share ever more proportion of the national economic growth and gradually become leading industries.
5. In the high degree processing and assembly-type-oriented industrial stage, due to the large number of high-tech applications, traditional industries are transformed. Some industries such as machinery, fine chemicals, petrochemicals, robotics, computer, aircraft manufacturing, aerospace, automotive, machine, and so on would have higher precision technical requirements. The high-value-added and assembly-type heavy industry would develop rapidly and promote the national economic development and become the driving force. They share a large proportion of the GNP and have a quick growth at the same time, and they become the leading industries of the national economy.
6. In the tertiary industry-oriented stage, the pace of the secondary industry development slows down, the proportion declines, especially in traditional industries; however, the inner new industries and high-tech industries show still more rapid development. The internal structure of the whole secondary industry rapidly changes, and the proportion is no longer a dominant one. The tertiary industry (including services, transport, tourism, commerce, real estate, finance, and insurance industry) speed up the pace of development significantly, and it has the major or larger share in the GNP and becomes the leading industry in the national economy.
7. In the information industry-led phase, the information industry develops by leaps and bounds. Particularly, the information superhighway construction and the popularity of the Internet promote the rapid development of information industry.

During this period, the information industry becomes a pillar industry and the leading industry of the national economy. It is also called the stage of postindustrialization.

2.2.3 Internal Changes of Three Industries

The evolution of industrial structure develops along with the direction of the primary industry-oriented to the second industry-oriented, and then to the tertiary sector-oriented industries.

In the primary industry, the industrial structure transforms from the extensive agriculture with the lower technology level to intensive agriculture with the high technical requirements, then develops to the higher levels of green agriculture, eco-agriculture such as the biological, environmental, biochemical, and ecological; the industrial structure also transforms from the planting agriculture to animal husbandry agriculture, from the wild-type agricultural to the development of plants agriculture.

In the secondary industry, the evolution of industrial structure develops to the light textile industry, heavy-chemical industry, and processing heavy industry. From the point of changing resources structure, the industrial structure evolves from labor-intensive industries to the capital-intensive industries and the knowledge-intensive industries (including technology). From the perspective of market-oriented industries, the industrial structure moves from the closed-type to the import-substitution type, the export-oriented type, and then to the direction of the market globalization.

In the tertiary industry, the industrial structure moves from the traditional service sector to the diversified service sector, the modern service industry, the information industry, and then in the direction of knowledge-based industries.

2.2.4 Order of the Evolution of Industrial Structure

The industrial structure develops from the low level to the high level. Although industrial structure is a long-term process, every stage of the development process can be shortened. From the perspective of evolution, the latter stages of the industrial development are based on full development of the previous stages. Only when the labor productivity in the primary industry developed fully, then the textile industry of the secondary industry can develop well. The development of the secondary industry is based on the substantial increase of the labor productivity in the primary industry; and the development of the processing and assembly-type heavy industry is based on the development of basic industries such as the raw materials, fuel, power, and so on. Likewise, rapid development of the secondary industry provides a mature and solid foundation to develop the tertiary industry. The advancing development of industrial structure will accelerate the development of the economy of a country. However, the development of industrial structure has also certain marginal effects. For example, South Korea's tertiary industry developed at a very rapid speed, and has a larger share. As early as 1962, the tertiary industry reached 45.6%; in 1970 the proportion of tertiary industry's output value was more than 50% in South Korea; by 1989, the proportion of the tertiary industry had risen to 59.4% which had already exceeded that of Japan, West Germany, France, Britain and of other countries in the same period. South Korea's premature tertiary industry was dominated by the traditional flow industry and service-oriented industry. However, the modern service industries such as finance, telecommunications, and information industry were not yet ripe, and the modernization level was not high. This kind of development along with the low modernization of the tertiary industry had an impact on South Korea's future economic development, which has been gradually revealed now.

2.3 Empirical Analysis of the Typical Countries

Here we take the evolution of industrial structure of United States, United Kingdom, Germany, France, Japan, and other countries as examples to further verify whether they are consistent with the general trends in the evolution of industrial structure mentioned above.

Yoshio Ando, the Japanese scholar in his *Modern History of Japan's Economy*, collected data of the above countries about the labor force distribution structure in three industries and the national income distribution structure in three industries from 1870s to 1970s. They reflected general trends in the evolution of industrial structure. The data are shown as Tables 2.5 and 2.6.

Table 2.5 Labor Force Distribution Structure (%) in Three Industries

		1870S	1880S	1890S	EARLY 20TH CENTURY	1920S	1940S	1950S	1960S	1970S
		[1872]	[1878]	[1897]		[1920]	[1936]	[1958]	[1963]	[1971]
Japan	Primary industry	85	78	72		55	45	37	29	16
	Secondary industry	5	9	13		22	24	26	31	35
	Tertiary industry	10	13	15		23	31	31	40	49
		[1870]	[1880]	[1890]	[1900]	[1920]	[1940]	[1950]	[1960]	[1971]
USA	Primary industry	50	50	42	37	27	17	12	7	4
	Secondary industry	25	25	28	30	34	31	35	34	31
	Tertiary industry	25	25	30	33	49	52	53	59	65
			[1881]	[1891]	[1901]	[1921]	[1938]	[1951]	[1966]	[1971]
UK	Primary industry		13	11	9	7	6	5	3	2
	Secondary industry		50	49	47	50	46	47	45	40
	Tertiary industry		37	40	44	43	48	48	52	58
			[1882]	[1895]	[1907]	[1925]	[1939]	[1950]	[1963]	[1971]
Germany	Primary industry		42	36	34	30	27	23	12	8
	secondary industry		36	39	40	42	41	44	48	48
	Tertiary industry		22	25	26	28	32	33	40	44
		[1866]			[1901]	[1921]	[1946]		[1962]	[1971]
France	Primary industry	43			33	29	21		20	13
	Secondary industry	38			42	36	35		37	39
	Tertiary industry	19			25	35	44		43	48

Source: Yang Zhi, *Introduction to Industrial Economics*, p. 41.

Table 2.6 Three Industries Structure Distribution of Gross Domestic Product (%)

COUNTRY	YEAR	PRIMARY INDUSTRY	SECONDARY INDUSTRY	TERTIARY INDUSTRY	P.S.
Japan	1895	42.7	18.2	39.1	NDP: price in the same year
	1925	28.1	30.0	41.9	NDP: price in the same year
	1963–1967	9.7	40.8	49.5	NDP: price in the same year
	1971	6	45.0	49.0	GNP: price in the same year
United States	1889–1899	17.9	44.1	38.0	NI: price in the same year
	1919–1929	11.2	41.3	47.5	GNP: price in 1929
	1963–1967	3.3	37.2	59.5	GDP: price in the same year
	1971	3.0	37.0	60.0	GDP: price in the same year
United Kingdom	1907	6.4	38.9	54.7	GDP: price in the same year
	1924	4.4	55.0	40.6	GDP: price in the same year
	1963–1967	3.4	46.3	50.3	GDP: price in the same year
	1971	3.0	38.0	46.0	GDP: price in the same year
Germany[a]	1850–1859	44.8	22.0	33.2	NDP: price in 1931
	1935–1938	16.2	50.3	21.5	NDP: price in 1931
	1963	5.0	54.0	29.0	GDP: price in the same year
	1971	3.0	53.0	33.0	GDP: price in the same year
France	1896	25.0	46.2	28.8	GDP: price in 1954
	1963	8.4	51.0	40.6	GDP: price in 1954
	1971	6	46.0	46.0	GDP: price in the same year

Source: Yang Zhi, *Introduction to Industrial Economics*, page 45.
Note: GDP, gross domestic product; GNP, gross national product; NDP, value of the net domestic product; NI, national income.
[a] Germany includes only the Federal Republic of Germany.

From Table 2.5, we find that with social and economic development, the employment proportion of the primary industry has declined drastically. In the United States it decreased to 46% from 50% in 1870 to 4% in 1971 in 100 years; in the United Kingdom it declined to 11% from 13% in 1881 to 2% in 1971; in Germany it decreased to 34% from 42% in 1882 to 8% in 1971; in France it decreased to 30% from 43% in 1866 to 13% in 1971; and in Japan it decreased to 69% from 85% in 1872 to 16% in 1971. The decrease of the employment proportion in primary industry is not only significant but also continuous, and the proportion is not increasing again.

The changes of the employment proportion in the secondary industry are different along with relative stages of the economic development in the above countries. United Kingdom and France were the mature industrialized countries; the employment proportion of secondary industry reached the highest value from the beginning of the study period, and then slowly declined. In Germany and Japan, the industrialization process was relatively late, the employment proportion of the secondary industry was at a lower level at the beginning of the study period, but it was in a rising trend over the study period. The employment proportion of the secondary industry in the United States was at a lower level at the beginning of the study period and reached the highest value in the 1950s and then had shown a downward trend. In particular, in the United Kingdom it dropped by 10% from 1881 to 1971 in nearly 100 years, which was a smaller decline; in France it was in high oscillation (maintained at about 40%) in the early 20th century to the 1930s, and then slowly declined. Just looking at the beginning of the study period and the end of the study period, the labor force proportion of the secondary industry changed only 1% in the two time-points; Germany had shown a rise from 36% to 48%; and in Japan it rose by 30% from 5% to 35%, and did not have any downward trend; in the United States it reached the highest value (35%) in the 1950s and then declined.

The employment proportion of the tertiary industry has been increasing with the economic development. In particular, in the United States it increased 40% from 25% in 1870 to 65% in 1971; in the United Kingdom it increased 21% in the past 100 years from 37% to 58%; in Germany it rose 22% from 22% to 44%; in France it increased 29% from 19% in 1866 to 48% in 1971; and in Japan it rose 39% from 10% to 49%.

From Table 2.6, we can see that with the economic development, industrial output value of the primary industry in proportion to the GDP was declining. In particular, in the United States it decreased 14.9% from 17.9% between 1889 and 1899 to 3% in 1971; in Britain it decreased 3.4% from 6.4% in 1907 to 3% in 1971; in Germany it decreased 41.8% from 44.8% between 1850 and 1859 to 3.0% in 1971; in France it decreased 69% from 75% in 1986 to 6% in 1971; and in Japan it decreased 36.7% from 42.7% in 1895 to 6 % in 1971.

The changing trend of the secondary industry in proportion to the industrial output value is as follows: in the initial stage of industrialization and the acceleration phase in one country, the proportion increased rapidly. When it was in a mature stage of industrialization, the proportion of its output tended to rise slowly and then stagnated and declined. In particular, in the United States it declined slowly from the beginning of the study period 1889–1899 from 44.1% to 36.0%; in Britain it slightly increased in the early 20th century, and then reached the highest value of 55% in the 1920s, and declined to 36.0% until 1971; in Germany it rose from 22% in the 1950s to 54.0% in the 1960s, and reached the highest value and then declined slightly to 53.0% in 1971; in France it showed a continued increasing trend in the study period from 46.2% in 1896 to 52.0% in 1971 rising in a small range; and in Japan it rose from 18.2% in 1895 to 44.0% in 1971, which was in a sustained increase stage.

The tertiary industry output value in the form of the study period did not show as obviously as in the case of the primary industry and the secondary industry. In particular, the proportion of the tertiary industrial output in the United States continued to rise from 38.0% to 60.0% with an increase of 22%; in the United Kingdom it declined from 54.7% in 1907 to 40.6% in 1924, and then rose to 50.3% from 1963 to 1967; in the initial study period, the proportion of the tertiary industrial output in Germany was 33.2% and at the end of the study it was 33.0%, which experienced a declining stage; in France it continued to show an increasing trend from 28.8% in 1896 to 46.0% in 1971 with an increase of 21.2%; and in Japan it rose from 39.1% in 1895 to 49.05% in 1971.

Until the 1970s, Yoshio Ando's data reflected the process of a number of developed countries reaching industrialization; from the data of several countries for the proportion of the output value in the secondary industry we can find that there is an obvious decreasing trend in the United States, United Kingdom, and France. However, the proportion of the secondary industry output value in Japan and Germany was still in the high-level maintenance phase or rising phase. The proportion of output value in the tertiary industry was also in fluctuation. In order to further reveal the new trend of the evolution of industrial structure in the developed countries since the 1970s, we had collected relevant data shown in Tables 2.7 and 2.8.

From Table 2.7 we can see that the employment proportion in the primary industry showed a continual decreasing trend just similar to the data in Table 2.5. In particular, in the United States it decreased 1.53% from 4.17% between 1970 and 1975 to 2.64% between 1997 and 1999; in the United Kingdom it decreased 1.3% from 2.97% between 1970 and 1975 to 1.67% between 1997 and 1999; in Germany

Table 2.7 Distribution of Three Industrial Structures of the Employment Population (%)

COUNTRY	YEAR	PRIMARY INDUSTRY	SECONDARY INDUSTRY	TERTIARY INDUSTRY
Japan	1970–1975	15.02	35.78	49.02
	1976–1980	11.31	35.58	52.90
	1981–1985	9.37	35.08	55.23
	1986–1990	7.91	34.01	58.04
	1991–1995	6.67	31.29	62.04
	1997–1999	5.38	32.38	
United States	1970–1975	4.17	33.57	65.36
	1976–1980	3.72	30.92	67.40
	1981–1985	3.31	29.29	69.87
	1986–1990	2.95	27.18	72.22
	1991–1995	2.89	24.89	73.64
	1997–1999	2.64	23.71	
United Kingdom	1970–1975	2.97	42.74	54.29
	1976–1980	2.67	38.70	58.61
	1981–1985	2.58	33.54	63.22
	1986–1990	2.34	29.56	66.89
	1991–1995	2.18	27.40	67.67
	1997–1999	1.67	26.33	71.61
Germany	1970–1975	8.00	47.65	44.35
	1976–1980	6.14	44.31	49.35
	1981–1985	5.02	42.18	52.80
	1986–1990	4.13	40.15	55.72
	1991–1995	3.37	37.93	58.64
	1997–1999	2.87	33.85	63.19
France	1970–1975	12.78	39.02	48.20
	1976–1980	9.29	36.90	53.81
	1981–1985	7.79	33.35	58.86
	1986–1990	6.22	30.27	63.51
	1991	5.33	29.11	65.56

Source: *The International Statistical Yearbook*, 1995, 2001.

Note: The period data is that of the average of the first and the previous year. Due to the restrictions of statistics, some employees who were in the unknown sectors were not divided into the three industries, so the total proportion of the labor force in the three industries will be less than 1.

it decreased 5.13% from 8.00% between 1970 and 1975 to 2.87% between 1997 and 1999; in France it decreased to 7.45% from 12.78% between 1970 and 1975 to 5.33% in 1991; and in Japan it decreased to 9.64% from 15.02% between 1970 and 1975 to 5.38% between 1997 and 1999.

Table 2.8 Distribution of Three Industrial Structure of the Gross Domestic Product (%)

COUNTRY	YEAR	PRIMARY INDUSTRY	SECONDARY INDUSTRY	TERTIARY INDUSTRY
Japan	1970–1975	6.15	47.15	46.65
	1976–1980	4.75	44.65	50.60
	1981–1985	3.55	44.25	52.15
	1986–1990	2.95	44.30	52.80
	1991–1995	2.20	41.70	56.10
	1997–1998	1.70	36.65	61.65
United States	1970–1975	3.30	36.95	59.70
	1976–1980	3.00	36.25	60.75
	1981–1985	2.65	34.70	62.65
	1986–1990	2.10	30.50	67.85
	1990–1995	1.80	27.45	70.80
	1997	1.70	26.20	72.00
United Kingdom	1970–1975	2.73	42.60	54.67
	1976–1980	2.54	42.00	55.47
	1981–1985	2.12	41.77	56.11
	1986–1990	1.95	36.41	61.65
	1991–1995	1.72	31.17	67.11
	1997–1999	1.15	26.20	72.55
Germany	1970–1975	3.38	52.71	43.92
	1976–1980	2.71	49.03	48.28
	1981–1985	2.14	46.16	51.70
	1986–1991	1.69	43.94	54.37
	1991–1995	1.32	36.30	62.39
	1997–1999	1.15	28.75	70.15
France	1977–1980	5.12	39.39	55.49
	1981–1985	4.59	36.04	49.38
	1986–1990	4.13	33.96	61.92
	1991–1995	3.24	28.61	68.16
	1997–1998	2.85	23.45	73.70

Source: *The International Statistical Yearbook*, 1995, 2001.
Note: The period data is that of the average of the first and the previous year.

The employment percentage in the secondary industry also showed a significant decline excluding exceptional individuals. In particular, in the United States it decreased by 9.86% from 33.57% between 1970 and 1975 to 23.71% between 1997 and 1999; in the United Kingdom decreased by 16.41% from 42.74% between 1970 and 1975 to 26.33% between 1997 and 1999; in Germany it decreased by 13.8% from 47.65% between 1970 and 1975 to 33.85% between 1997 and 1999; in

France it decreased by 9.91% from 39.02% between 1970 and 1975 to 29.11% in 1991; and in Japan it decreased 4.49% from 35.78% between 1970 and 1975 to 31.29% between 1991 and 1995, during the period 1997 and 1999 it had a slight increase of 32.38%.

The employment proportion in tertiary industry continued an increasing trend. In particular, in the United States it increased by 8.28% from 65.36% between 1976 and 1980 to 73.64% between 1997 and 1999; in the United Kingdom it increased by 18.84% from 54.29% between 1970 and 1975 to 71.61% between 1997 and 1999; in Germany it increased by 17.32% from 44.35% between 1970 and 1975 to 63.19% between 1997 and 1999; in France it increased by 17.36% from 48.20% between 1970 and 1975 to 65.56% in 1991; and in Japan it increased by 13.02% from 49.02% between 1970 and 1975 to 62.04% between 1991 and 1995.

Comparing Table 2.7 with Table 2.5, we can derive the following conclusions: the employment proportion of secondary industry in a few developed countries had continued the decreasing trend since the 1970s, while the employment proportion in the tertiary industry had continued to rise (the range was much higher than that in Table 2.5). This showed that several developed countries had developed from the mature stage of industrialization to the "postindustrialization" stage.

From Table 2.8, we can see that the proportion of the primary industry output value declined continuously. In particular, in the United States it decreased by 1.60% from 3.30% between 1970 and 1975 to 1.70% in 1997; in the United Kingdom it decreased by 1.58% from 2.73% between 1970 and 1975 to 1.15% between 1997 and 1999; in Germany it decreased by 2.23% from 3.38% between 1970 and 1975 to 1.15% between 1997 and 1999; in France it decreased by 2.27% from 5.12% between 1977 and 1980 to 2.85% between 1997 and 1998; and in Japan it decreased by 4.45% from 6.15% between 1970 and 1978 to 1.70% between 1997 and 1999.

With the emergence of the service economy, the proportional output value of the secondary industry also showed continuous decreasing trend. In particular, in the United States it decreased 10.75% from 36.95% between 1970 and 1975 to 26.20% in 1997; in the United Kingdom it declined by 20% from 46.20% between 1970 and 1975 to 26.20% between 1997 and 1999; in Germany it decreased by 23.96% from 52.71% between 1970 and 1975 to 28.75% between 1997 and

1999; in France it decreased by 15.94 points from 39.39% between 1977 and 1980 to 23.45% between 1997 and 1998; and in Japan it decreased by 10.5% from 47.15% between 1970 and 1978 to 36.65% between 1997 and 1999.

The proportion of output value in the tertiary industry has continued to rise. In particular, in the United States it increased by 12.3% from 59.70% between 1970 and 1975 to 72.00% in 1997; in the United Kingdom it increased by 17.88% from 54.67% between 1970 and 1975 to 72.55% between 1997 and 1999; in Germany it increased by 26.23% from 43.92% between 1970 and 1975 to 70.15% between 1997 and 1999; in France it increased by 18.217% from 55.49% between 1977 and 1980 to 73.70% between 1997 and 1998; and in Japan it increased by 15% from 46.65% between 1970 and 1978 to 61.65% between 1997 and 1999.

Comparing Table 2.8 with Table 2.6, we find that the output value in the secondary industry showed continuous decreasing trend while the output value proportion in the tertiary industry had shown a sustained increasing trend during the study period shown in Table 2.6.

3

Rationalization of Industrial Structure

The theory of industrial structure optimization is an important component of the theory of industrial structure and an important economic theory for a country to achieve sustainable, stable, and rapid development. Industrial structure optimization refers to the process of promoting the development of industrial structure rationalization and advancement. The former is to adjust the inconsistent industrial structure, mainly according to the technical and economic ratio of industry linkage, and to promote a balanced development among various industries in the national economy; the latter is to accelerate the evolution of industrial structure advancement by innovation, adhering to the evolution law of industrial structure. The process of industrial structure optimization is to adjust the supply and demand structure, which affects changes of industrial structure by the government's industrial policies to realize optimal allocation and reallocation of resources and push forward the development of industrial structure rationalization and advancement. This chapter focuses on the meaning and criteria of the industrial structure rationalization, as well as other relevant issues. Chapter 4 will elaborate on industrial structure advancement.

3.1 Meaning and Content of Industrial Structure Rationalization

3.1.1 Meaning of Industrial Structure Rationalization

Industrial structure advancement is based on industrial structure rationalization, without which it will be difficult to make advancement. Industrial structure rationalization refers to the strengthening of coordinating ability and the improvement of linkage levels between industries as well as inter-industry linkages. It is a dynamic process to promote the dynamic equilibrium of industry structure. Therefore, the major issues it needs to tackle include coordinating development between three industries and sectors within the industry, the mutual adaptation of supply and demand structure, and the full play of the effect of industrial structure. The industrial structure rationalization requires, at a certain stage of economic development, adjustment of relevant variables of the industrial structure, which is not ideal initially, according to the consumer demand and resources, thus straightening out the structure and allowing resources to distribute rationally and use effectively among industries. The key point of measuring the rationalization of the industrial structure is to determine whether an overall ability (different from the sum of the abilities of all industries) arising from their internal interaction exists among industries, that is, the quality of industrial aggregation.

The quality of industrial aggregation is the essence of industrial structure rationalization. In terms of system theory, there is an organic linkage among industries, in which the interaction among industries will produce an overall ability that is different from the sum of the abilities of all industries, namely, the aggregation quality of industrial structure. The more coordinated the interaction among industries, the higher the aggregation quality of industrial structure. Therefore, the industrial structure is reasonable and vice versa. Regarding the aggregation quality of industrial structure as the essence of rational industrial structure and making the degree of industrial coordination which increases the aggregation quality as a judge of rational industrial structure are based on the structure effects that the industrial structure has on the economic growth and benefits. Detailed analyses are as follows:

1. *Aggregation quality reflects the basic characteristics of the industrial structure.* On the surface, the industrial structure seems to be a unification of the production relation on production scale and the mode of connection among industries. However, this is only superficial and does not reflect the basic characteristics of the industrial structure. The essence of industrial structure is the technical and economic ties among industries shaping in economic activities. As an organic, wholly constituted by various industrial sectors in accordance with a certain connection mode and combination method, the industrial

structure adheres to the principle that the system as a whole is more than the sum of its parts and could present an overall benefit or "additional capacity" which various industries cannot have. Such a kind of overall effect, which is different from the simple arithmetic sum of the functions of all industries, is the basic feature of industrial structure.

2. *Aggregation quality reflects the overall quality of industrial structure.* In a system, structure describes the overall nature of the system internally, whereas a function depicts it externally. As the quality and the environment of elements of a system are different, the phenomenon that two systems have the same structure but different functions will occur. Therefore, it is difficult to evaluate the overall quality of an industrial structure simply by inter-industry linkages and ratios. Sometimes the combination of industrial structure is very effective in some countries, but not appropriate for others. This is the representation of the phenomenon "the same structure, different functions" in the system of industrial structure, owing to different industrial qualities and external environment between industry starters and latecomers. This phenomenon is particularly evident in the society of dual economy where there is a wide gap among industrial qualities. Hence, to evaluate the overall quality of the industrial structure, we have to take into consideration not only the connection modes and ratios of various industries, but also the level of the aggregation quality.
3. *Aggregation quality reflects the system efficiency of industrial structure.* Leading industries contribute significantly in a given time and a given industrial structure. However, the conversion efficiency of the industrial structure system to resources depends not only on the system's strongest element (the development of leading industries), but also on the weakest link (the conversion efficiency of the bottleneck industry). Of course, as a result of the partial substitution of resources as well as the flexibility of the system, the restriction of the conversion efficiency of the bottleneck industry can be partly alleviated through the coordination mechanism. If the supply and demand structure is in conflict, the industrial structure

should adapt to the normal demand changing provided that its aggregation quality is high. Therefore, an evaluation of the overall efficiency of the industrial structure system should be carried out by the aggregation quality of the structure, which reflects the integrated efficiency well, rather than by the more advanced industries.

3.1.2 Content of Industrial Structure Rationalization

We can see that coordination is central to industrial structure rationalization. Coordination is not only referring to the absolute balance among industries, but also to the strong complementary and harmonious relation among industries and the ability to convert each other. Only by strengthening the coordinating relations among industries, can we improve the aggregation quality of the structure, which in turn enhances the overall effects of the industrial structure. The coordination of various relations among industries is involved in the coordination of industrial structure, such as production, technology, distribution, and interests. We can observe and analyze whether there is coordination among industries from the following aspects.

1. *Whether the industrial qualities are coordinated.* That is to see whether there is a gap in technical level and a strong difference in labor productivity among related industries. If it does, there will be a massive friction. We can use comparative labor productivity, that is, the share of national income to the share of the labor force of a certain industrial sector, to measure generally the degree of coordination. In general, if the figures of comparative labor productivity are concentrative and have different levels, it means the qualities of various industries are coordinated and vice versa.
2. *Whether the connection modes among industries are coordinated.* There is a relation between input and output that is able to demonstrate the interdependence and mutual influence among industries. The mode of connection among coordinated industries features two basic characteristics. First, to serve each other, that is, various sectors provide help for each

other on the basis of input–output relations. For example, the development of agriculture provides raw materials and labor forces for the industry, which in turn provides the development of agriculture machinery equipment and technology; second, to promote industry development in each other, which means the development of one industry cannot be gained at the expense of undermining the other. If the abovementioned two aspects exist, the connection modes among industries are coordinated; if they do not, then they are uncoordinated.

3. *Whether the relative status among industries are coordinating.* In a certain stage of economic development, the economic function and the corresponding growth rate of various industries are different. Therefore, their status in the industrial structure differs, which form orderly permutation and combination of various industries. The coordination of the relative status of various industries refer to the permutation and combination, which have rich levels, and the primary industry, the secondary industry as well as the development priorities are obvious and appropriate and vice versa.
4. *Whether the supply adjusts to the demand.* On the premise of normal changes in demand, the coordination of the industrial structure will enable it to have strong adaptability and resilience. Namely, by its own restructuring to adapt to the new changes in demand, the disparities between supply and demand, both in quantity and in structure, gradually narrows and the contradictions gradually weaken too. On the other hand, if supply cannot respond to the normal change in demand, resulting in long-term imbalance between supply and demand, the industrial structure is uncoordinated.

In short, the symbol of industrial structure rationalization is as follows: to make full and effective use of native human, material, and financial resources as well as the benefits of the international division of labor to achieve coordinated development of all sectors of the national economy, to smoothen the process of social production, exchange, and distribution, favoring development through the expansion of social reproduction; to promote continued steady growth of the national economy and realization of social demand; finally, to achieve a benign cycle of population, resources, and environment.

3.2 Judging Benchmark for Industrial Structure Rationalization

3.2.1 Benchmark of International Standard Structure

Chenery's study of "standard industrial structure" reflects a certain law of evolution of industrial structure as its comprehensive description in most countries, and can be used as a frame of reference to judge whether a specific system of the industrial structure is reasonable. However, in the development of the industrial system, it is difficult to form a unified model of development and the industrial structure because of the differences of their space-time environments, with each having its own unique path of development. Therefore, it is not possible to use a standard model to judge whether industrial structures of various countries in different periods are reasonable. We will take the relationship between the country of industrial starters and the latecomers and between the structure of a big country and a small country as examples to illustrate the fact that the industrial structure system can form a track which is different from the "standard structure" as a result of different space-time environments.

1. Depending on the period and background characteristics of industrialization process, the countries can be divided into two types: industrial starters and latecomers. The major difference of this division is the background characteristic at the time when the said country started the process of industrialization, which is the condition of economic development at the beginning of industrialization in the country. The fundamental difference between starters and the latecomers is that the former starts the process of industrialization after elementary completion of its economic development; while the latter in the absence of economic development, following the industrialization model "first to start the economic take-off, then to complete the task of development." Consequently, the former runs a smooth track of industrial structure, while the latter runs that track aslant and has accelerated characteristics in some conditions. From the results of a variety of factors, the

latter reveals the features of "compression" evolution in terms of the running speed of industrial structure. The so-called "compressed" evolution refers to the speed of change of industrial structure, or transformation speed that accelerates in a certain stage. It generally takes a fairly long time for the starters to complete the mission of structural transformation, which can be done by the latecomers in a relatively short period of time by "compression."

2. According to the scale of domestic market and resource endowment, different countries can be divided into "great power" and "small country" structure. In terms of the development space, industrial structures of different countries have the difference of "great power" and "small country" structure. The two types are determined by the scale of domestic market and adequacy of resource endowment, which affect the industrial structures of the respective countries. The industrial structure system with larger domestic market and more abundant resources is called "great power structure," while "small power structure" refers to the industrial structure system, which has small domestic market or lack of resources. As the "great power" and "small country" face different conditions in building their industrial structures, their growth tracks are different under common circumstances.

According to studies of Kuznets and Chenery, the "great power" and "small country" have some differences in their industrial structures. "Great power" could enter the stage of rapid changes of structure at lower income levels relative to the "small country" structure. The "great power" structure has a larger market scale which can achieve large-scale production at a lower level, and adapt itself to manufacture in massive production systems, so it can begin development when the per capita income is lower. However, the smaller market scale does not fit manufacturing in massive production systems. Its manufacturing sector can only be developed with foreign markets in good international trade environment. It needs more actual strength to open foreign markets than winning domestic markets. This proves that the rapid growth in the manufacturing sector of the small country's structure is later than that of the great powers.

3.2.2 Benchmark of Demand Structure

If we view the industrial structure as a resource conversion system, it can be expressed in the following steps: elements supply structure—the industrial structure system—the market demand structure. In the formation of this system, every part has different characteristics owing to different elements and objective functions. For example, due to the effect of higher income levels and demand preferences, the demand structure of the market has characteristics of changeability when the overall consumption level increases whereas the industrial structure shows a strong rigidity as a result of the restrictions of precipitation of the production equipment. The benchmark of demand structure is to determine whether the industrial structure is rational by the degree of adaptation between the demand structure and the industrial structure. The higher the degree, the more rational the industrial structure and vice versa.

There are many factors affecting the rationality of the demand structure; among others, the first and the foremost is the proportion between the intermediate demand and final demand which also determines the production relation between industries producing intermediate goods and those producing final products. In terms of the trend of structural transformation in the industrialization process of various countries, the technological progress within a department, the increased mechanization level of the production process, and the deepening of reprocessing make the larger share of the total output into intermediate inputs, rather than the ultimate consumer goods. This means that the expansion of the scope of labor division and the increasing level of specialization has resulted in a more detailed structure of sectors and a larger transaction scale. As for the social production system which operates conventionally, the total increase in intermediate demand depends on changes of the total final demand and the structure and the economic growth of production sector of intermediate goods depends on the expansion of social aggregation demand. The Engel coefficient has an impact on the expansion of social aggregation demand and indirectly activates the rapid growth of production sector of intermediate goods by stimulating investment in equipment. This course of ultimate demand to aggregation demand and then to the intermediate demand reflects the dynamic characteristics of the growth of sectors in the conventional process of industrialization. For the social production system which does not operate conventionally, overexpansion of the intermediate demand possibly results from the too large material consumption coefficient and the abnormal structure of sectors; therefore, this does not mean the actual demand for the intermediate goods as a result of effective economic growth is increasing. Under such circumstances, to increase the production of intermediate inputs in accordance with the demand will only aggravate the imbalance of the industrial structure system. This will be followed by the personal consumption structure. Consumption can be regarded as the function of income, $C = f(y)$. The propensity to consume is divided into average propensity to consume and the marginal propensity to consume. The former refers to the ratio of total consumption to total income (c/y); the latter is the ratio of incremental consumption to incremental income ($\Delta c/\Delta y$). Changes in the industrial structure and propensity to consume are closely related. A large number of statistical analyses show that the changes in demand structure and changes in the industrial structure are parallel. Petty–Clark theorem is an experiential estimation on transferring of resources among industries caused by difference of comparative advantages. The increase of income levels stimulates the division of demand systems and industries and causes the difference of interests among industries and sectors, which lead to the transfer of resources from the field of agricultural production—producing basic survival material—to the integrated production fields which have a higher degree of processing. Adapting to the shift of industry focus, the service industry gradually becomes independent from agriculture and industry and becomes a separate sector. The formation of three major industries is the result of the fact that the demand structure and individual consumption structure changes in accordance with the income levels, as well as the result of the ratio of consumption to investment and the investment structure. The ratio of consumption to investment in the final demand has an obvious impact on the industrial structure and determines the proportional relation between industries of consumption material and those of capital material. Furthermore, investment can bring into being a new production capacity. Investment distribution in various industrial sectors is the direct result of changing the existing industrial structure. Changes in investment structure are subject to changes of production process, production technology as well as changes in the organic composition of capital.

As mentioned above, the malformed industrial structure means that it has a serious departure from the demand structure. In this sense, this benchmark is rational. However, to judge the rationalization of the industrial structure purely using this benchmark is one-sided. We have to first make sure whether the demand is normal, in the premise of which that judgment can be carried out. In case of abnormal demand, the gap between supply and demand is normal. If the industrial structure changes in order to adapt to the abnormal demand, such change is actually irrational.

3.2.3 Benchmark of Balanced Ratio among Industries

As mentioned above, we consider whether the ratios among industries are balanced as a criterion to determine whether the industrial structure is reasonable. In theory, economic growth is achieved on the basis of coordinated development of various industries. Balanced ratios among industries are basic conditions for economic growth. Whether the relations among industries are coordinated is reflected in the fact whether the ratios among them are reasonable. A researcher at the Institute of Economic Research, Shanghai Academy of Social Sciences, Zhou Zhenhua had once put forward the method of using "the degree of balance in the ratio" to measure the degree of equilibrium and coordination of the industrial structure system with the help of input–output analysis.

Suppose the actual output of the industry i is X_i' $(i = 1,2,\ldots,n)$, the output of meeting all the demand is X_i $(i = 1, 2, \ldots, n)$, then the imbalance volume of the industry i is

$$S_i = X_i' - X_i$$

Defining the imbalance coefficient as

$$K_i = \frac{S_i}{X_i}$$

the degree of imbalance in the ratio of the whole system of the industrial structure is

$$a = \sum_{i=1}^{n} |K_i|$$

The degree of imbalance in the ratio of the system is given by

$$b = \frac{1}{1+a} = \frac{1}{1+\sum_{i=1}^{n}\left|\frac{X_i' - X_i}{X_i}\right|}$$

The ranges of a and b are $(0, \infty)$ and $(0, 1)$, respectively.

However, this benchmark cannot be regarded as absolute. It is just reasonable to maintain such a balance of the industrial structure no matter when and where. In fact, under the circumstances of nonequilibrium economic growth, it is normal that the growth rates of the industries vary. Only a structural imbalance beyond the limit will lead to an improper functioning of the economy. This is the real irrationality in structure.

The industrial structure is a system for which the integrated features require the indivisibility of its components. In this system, production capacity and overall output levels will greatly weaken in the absence of a balanced proportion among industries. According to the "principle of casks," the full play of the function of the industrial structure system depends on the industry with the weakest output capacity rather than on one with the strongest output capacity. When an industrial structure system has bottleneck industries, its production capacity will be subject to the exertion of these bottleneck industries. Usually an industry with a small proportion of intermediate input and a large proportion of intermediate demand is called intermediate input industry. We analyze the development of the bottleneck industry in the system of industrial structure in accordance with this standard.

3.3 Comparison and Determination of Industrial Structure Rationalization

At present, when we identify and argue changes in the direction of the industrial structure and whether the industrial structure is rational, we usually adopt the following methods of analysis:

1. *The method of international comparison.* On the basis of Chenery's standard industrial structure, we can compare the industrial structure of a country with the standard one that

has the same GNP. The industrial structure is irrational if there is a big deviation. This method can only approximately judge whether the industrial structure is rational rather than determine it decisively.

2. *The method of shadow price analysis.* This is to find out whether the industrial structure is rational by the shadow price, which is different from the actual market price and is "the price that reflects the optimal use effect of resources calculated by the method of linear programming." In constrained optimization in economics, the shadow price is the change in the objective value of the optimal solution of an optimization problem obtained by relaxing the constraint by one unit, which is the marginal utility of relaxing the constraint, or equivalently the marginal cost of strengthening the constraint. According to the theory of western economics, when the marginal outputs of various products are equal indicating that resources are allocated reasonably, the supply and demand of a variety of products are balanced and industrial sectors have achieved the best combination. As a result, each sector's shadow price can be used to compare with the overall average shadow price. Its deviation can be taken to estimate whether the industrial structure is rational. The smaller the deviation, the more rational the industrial structure.
3. *The method of the judgment of demand.* This method is used to estimate whether the actual production capacity of various industries matches the corresponding demand for its products. If the two values are close, then the current industrial structure is rational.
4. *The method of the judgment of demand adaptability.* This method is to estimate whether the industrial structure can realize self-regulation along with the changes in the demand structure in order to adapt to the demand structure and achieve the goal of social production. The method is to calculate the income elasticity of demand and of production of each industrial product; if these two values of a certain industry are equal, then we say the industry has sufficient adaptability to social demand; if these two values of every industry are equally divided, the whole industrial structure is suited to the

structure of demand, then the industrial structure is reasonable.

5. *The method of judgment of structural effects.* This method is based on changes in the total output and total profits of the national economy, and caused by changes in the industrial structure, to measure whether the industrial structure moves toward the direction of rationalization. Specifically, if changes in the industrial structure lead to the relative increase of total output and total profits of the national economy, then it means that the industrial structure is on the road to rationalization; on the contrary, if changes in the industrial structure result in relative decline in total output and profits, then the industrial structure moves toward the direction of no rationalization. In a specific sense, the increase of macroeconomic benefits is from the high efficiency of the industrial structure. This is because a rigid industrial structure and low efficiency level of structure mean that there are a large number of inefficient industries and the stock transferring among industries is extremely slow. Under such a situation, the efficiency of resources utilization is very low. Without a change in this kind of industrial structure, it may raise the economic growth rate and the total profits. However, it will not improve economic benefits no matter how much more investments are made. Only by high efficiency of the industrial structure, which means a continual decline of the proportion of inefficient industries and a continual increase of the proportion of efficient ones, can we raise the level of macroeconomic benefits.

3.4 Adjustment of Industrial Structure Rationalization

3.4.1 Process of Adjustment

For the process of rationalizing the industrial structure adjustment, there are two main processes: one is that of continual adjustment and coordination among sectors and trades to balance it; the other is that of breaking the balance. Excluding special circumstances, the main reasons are that the balance is broken. First, the industrial structure adjusts following the changes of supply and demand structure; second, some changes occur in the supply abilities of some industries because of technological advance; thus the industrial structure needs some adjustments to adapt to the relatively unchanged supply and demand structure.

If the marginal revenue of the industrial structure decreases without major technical level changes in a short term, this is in accordance with the course of becoming a reasonable industrial structure from being an unreasonable one. This is due to the process of industrial restructuring as well as the narrowing process of the extent of structural distortion. With the industrial structure becoming more and more coordinated, the economic loss sustained from distortions of the industrial structure is gradually reducing. So the amount of revenue received by correcting the distortions will be less and less. However, studying the changes and development of the whole industrial structure in a longer period of time, we can see that the margins of continual restructuring resulting from technological progress do not follow the law of diminishing returns. The labor forces and material production factors required for meeting certain demands have been economized due to technological process. The productivity increases twofold. If each round of industrial restructuring due to technological progress is regarded as the "margin" of changes in the whole industrial structure, the marginal revenue does not diminish.

3.4.2 Mechanism and Drive of Adjustment

An irrational industrial structure develops toward the direction of rationalization. Its momentum is the existence of events in the process of restructuring. However, in different restructuring mechanisms, the representative forms of the restructuring drive are different. Industrial restructuring mechanism is the process of formation of a new industrial structure state by inputting some kinds of signals and energy to cause structural changes in accordance with the existing industrial structure. According to the nature of the input signals and the type of adjustment, theoretically, the mechanism could be divided into market and planned mechanisms.

1. *The market mechanism of industrial restructuring.* The market mechanism to a large extent is a self-adjustment process of the economic system, namely, the main economic body adapts the

industrial structure to the changes of demand structure as much as possible by reorganizing the productive resource and its flow among industrial sectors under the guidance of signals in the market. The structure of demand changes for various reasons. It destroys the original structure of supply and demand. The supply of certain products is larger than the demand, whereas the demand of other products is larger than the supply, which causes corresponding fluctuations in the prices of these products. When the price fluctuations are large enough to reach the critical point for inter-departmental transfer of productive resources (the proceeds after transfer = the transfer costs + opportunity cost), resources of the sector having a decline in their product prices will be transferred to the sectors having an increase in their product prices until it reaches a new point of balance between supply and demand. In the course of industrial restructuring, the signal of change in the structure is the market price; the driving force is the pursuit of increasing profits and avoiding losses by numerous scattered main economic bodies.

2. *The planned mechanism of industrial restructuring.* It is the process of regulation and control on economic system, namely, the government inputs some kind of signals to the economic system and directly allocates resources among industries in order to change the industrial structure. Depending on the existing state and the forecast of changes in the industrial structure, the government will promulgate instructions to the main economic bodies by vertical hierarchy and will adjust the supply and demand among industrial sectors with the overall objective of economic development. There are usually two types of these instructions: one is to request directly the production quantity of enterprises; the other is to adjust the allocation of asset increment among industries through changes in the investment plans of various departments to change the industrial structure. In such industrial restructuring, the signal of changes in the structure is the planned quantity or instructions of the government; the driving force is the government's pursuit of sustained, steady, and coordinated economic growth.

The two kinds of adjustment mechanisms have strengths and limitations of their own. The market mechanism is relatively accurate, safe, and sensitive, but has a relatively large cost and a long time-lag owing to after-adjustment; the planned mechanism has the advantage of before-adjustment, thus a smaller adjustment cost. However, it is not very accurate and has a larger market friction. Consequently, the use of one single regulator approach is difficult to achieve in the goal of rationalization of industrial structure. Only through an organic combination of the two mechanisms can the industrial structure be rationalized. At present, no country in the world adopts the single approach of market or planned adjustment. What is used is a combination of the two forms, which has different focuses.

4

Advancement of Industrial Structure

The advancement of industrial structure denotes the dynamic process in which the industrial structure develops from a low level toward a high level. It indicates the abandoning of the original industrial structure, which involves the gradual conversion of the dominating industry from the primary to the secondary and tertiary industries. And the industrial structure development shifts from being labor-intensive to capital-intensive, and then to tech-intensive successively; the industrial structure also changes from that of a low value added industry to one of high value added industry. The industries manufacturing primary commodities will be gradually be replaced by those manufacturing middle products and final products.

4.1 Dynamic Reasons of the Advancement of Industrial Structure

The market demand is the external factor of influence for the industrial structure advancement. To some extent, the industrial structure of a country is a resource conversion system. On the one hand, it absorbs various products from the outside world; on the other hand, it provides various products depending on the market. The change in market demand will necessarily cause a change in production and production structure and directly influence the development direction of the industrial structure. The market demand can be divided into consumption demand and investment demand according to usage. Consumption demand is the ultimate social demand. The income level of the consumers establishes the foundation of consumer demand structure, which reflects the change in industrial structure. Along with the increase of income level, the Engel coefficient decreases continuously and the consumer demand changes from basic subsistence goods to enduring consumption goods.

Petty–Clark theorem also believes that the increase of per capita income will stimulate different demand system and industry, which will cause positive differences between industries and sectors. In the end, resources will be forced to transfer from basic subsistence industry to comprehensive production realm with a higher degree of processing. In the market economy, most investment decisions are made by the enterprise. The enterprise, as a rational "economic person," decides the resources allocation according to the market demand signals; the enterprise also invests differently in different industrial directions, which changes the existing industrial structures and forms new ones.

Technological progress is the direct reason for the advancement of industrial structure. A certain technology level always relates to particular industrial structures. When innovation causes technological level increase in some sectors and the expansion of technology into other sectors, the change in industrial structure will be stimulated to push it to a higher level. Technological progress will promote the advancement of industrial structure in the following ways: (1) It renews and perfects the original production crafts and techniques, which makes production more efficient and causes industrial scale expansion and quality structural changes. (2) It expands the scope of useful resources. Usage of new technology lowers the requirement of resource quality, thus inducting more resources into the production realm, increasing the supply of resources, and making the transfer of resources between the three industries possible. (3) It makes large-scale production possible, and lowers labor costs and increases labor productivity. (4) It creates new demand, which makes part of potential market demand change into real market demand. For the industry whose output elasticity is more than 1, the improvement of technology of that industry results in lower cost for unit production and hence a lower price; the demand scale for the product of this industry will be more, attracting inflows of production factors and causing a change in the industrial structure.

Industry policies are important forces to push the advancement of industrial structure. In order to reduce the twists and turns of the process of advancement of industrial structure and overcome market failure, the government strengthens its understanding about the evolution of industrial structure and follows the market rules. With the help of industrial support, adjustment, technology, protection, organization, layout, and so on, the advancement of industrial structure will speed up and the upgrading of industrial structure will be realized in as short a period as possible, thus enhancing international competition capacity.

The industry policy is part of the national macro-policies, the principle of establishment of which includes the industrial structure and comparative advantages of the native country, the ability of the technological progress, the degree of market process, the availability of resources, the strategic key points of economic development, and the trend of the world industry development. The fundamental function of industry policies lies in the valid intervention of the government and indicates the direction of industrial development and opens the native industry realms in an orderly way and accelerates industrial technological progress.

Generally speaking, in order to catch up with the developed countries, industrialized nations usually adopt the strategy of locating industries in advanced economy or to develop the economy by several stages at the same time. They will struggle for the realization of industrialization in a short time and complete the conversion of three industries. Japan is a nation that adopted this kind of method to push forward the advancement of industrial structure and made it a success. After World War II, Japan's high economic growth rate drew worldwide attention.

Because of official intervention of the government, Japan resumed after several years and realized heavy industrialization from mid-1950s to 1970s. Just from the distribution of labor resources, the primary industry took up 50.8%, the secondary industry 22.3%, and the tertiary industry 26.8% in 1950; by 1970, the primary industry decreased from 33.4% to 17.4%, the secondary industry increased by 12.9% to 35.2%; the tertiary industry increased by 20.5% to 47.3%. In 1980, the situation of the labor force in the three industries was as follows: the primary industry declined to 10.8% and the secondary industry declined to 34.4% and the tertiary industry reached 54.8% increasing by 7.5%. This shows that Japan spent only over 40 years after World War II realizing the three conversions from light textile industry to infrastructure industry, to heavy processing industry, to service industry, and promoted the advancement of industrial structure quickly.

4.2 Functioning Mechanism of the Advancement of Industrial Structures

The advancement of industrial structure is realized by shifting the position of advantage among industries. The upgrading of industrial structure is the comprehensive result of changes in each industry, which is based on changes in each sector. From changes in each sector, the following dynamic process of "rising–extending–decelerating–shrinking" will be experienced. The rise of an industry is usually associated with the development of a new product. As public understanding about the advantages of new products grows, the demand for them will be enhanced day by day. In the meantime, innovation successfully will lower the cost of products significantly to make the industry expand quickly and get into a quick growth stage. However, when this kind of high-speed growth attains a certain critical point, a trend of decelerating growth will appear. After analyzing a large number of materials, Kuznets found this kind of phenomena in which the growth rate of industrial sectors will slow down, and believed that in the long run, there are certain natural rules. Just as he pointed out, the growth rate of most sectors of the total production did decrease in a period of time; thus, the quota they shared in the total production value of the whole country also decreased correspondingly. Therefore, although typically there is no deceleration of growth for the total production value and per capita production value, it is not typical of the economy in a country. The deceleration of growth of industry sectors is caused by the following reasons: technological progress happens at a slow rate; the slowly growing industry exerts a dampening effect on its growth and the industry whose growth is fast will inhibit its competition. Along with the growth of the industry, the proportion of funds available for expansion decreases. It is obvious that the trend of the growth rate will decelerate despite industrial sectors abiding by certain rules over the long history of the industry. These are the theoretical foundations and prior conditions for the analysis of changes in industrial structure upgrading.

Innovation is one of the main factors that influences the change in industrial sectors, reduces the cost of the products greatly and makes the industry grow quickly. However, when the price of the product declines to a degree where innovation does not make a difference, namely, the potential for innovation to reduce the cost has dried up, innovation will slow down. As a result, the industrial growth rate will decelerate. Since the development of any industrial sector depends on innovation and shows the rule for extending or shrinking, each industrial sector of a nation can ensure its relative position according to its status of innovation.

Kuznets' research shows that the industrial growth rate will change from high to moderate to slow along its growth path, becoming mature and scaling down the growth rate of the industry for a long time. We see industrial sectors with various rates of growth at any point of time. Generally speaking, high growth sectors are at comparative advantage because they continue the practice of innovation, take up a bigger share in the total market value, and prop up the growth of the whole economy. With economic development, the change in industrial structure shows the change in high growth dominating industries with new innovation and its proliferation, which is a continuous process of change. When the originally high growth industry decelerates as innovation slows down, it would be replaced by a new high growth industry. In the later developing process, the potentially high growth industry will convert again into the real high growth industry to replace the original high growth industry. Therefore, the change in industrial structure is realized through changes in the position of advantage of the industries.

The position of advantage of an industry can be measured by three standards, one of which is added value. The industry with a high added value has high advantage; the second one is the production value of an industry; those who have a larger production value are advantage industry; the third one is the industry connection effect. The industries are much affected by both raw material supply and ultimate demand. If they have larger backward and forward connection effects, then they are considered dominating industries. The added value standard emphasizes the amount of profit margin. The production value standard emphasizes the extension of production value scale, and the connection effect standard emphasizes the influence of industry. The increase in the added value, extension of production value scale, and the strengthening influence of industries all depend on innovation. It is innovation that causes change in added value, production value scale, and industry influence in different industries to push the advancement of industrial structure.

4.3 Judgment Standard of the Advancement of Industrial Structure

4.3.1 International Standard Structure of the Advancement of Industrial Structure

There are some regulations for the evolution of industrial structures. The discretion of advancement of a particular industrial structure system can be compared with the standard structure. Standard structure is the comprehensive description of the evolution of the advancement of industrial structure in a majority of nations. Generally, the method of statistical analysis will be used to induce characteristics of advancement of industrial structure in sample nations, on the basis of which some indicators can be obtained for a certain upgrading stage. Then these will be the "standard" structures for an industrial structure to evolve to.

Using the standard structure in the practical analysis for the advancement of industrial structure, Kuznets, Chenery, and Syrquin made great contributions. The setting up of the standard structure is to make use of statistical analysis and regression summary with the help of regression analysis of the main factors that influence sample nations or reflect the industrial structure. In practical applications, the standard structure based on production value or the labor force is often used.

4.3.1.1 Standard Structure That Takes Production Values as Indicators When the industrial structure system is viewed from the angle of systems, the output of the system is the output of industrial structure. The constitution of these output and the form of mutual relations in the industrial structure are important. So the output structure is an important angle from which to observe industrial structure. The production value structure is a phenomenon of certain price systems, and the industrial structure is analyzed and researched using this structure.

Since Kuznets, many economists engaged in the research of industrial economics began to pay more attention to the research of the relation between the production value structure and the advancement of industrial structure. Many standard structures also take production

Table 4.1 Pattern according to Chenery, Elkington, and Sims

	BASIC LEVEL OF PER CAPITA GDP[a] (%)							
INDUSTRY	US$ 100	US$ 200	US$ 300	US$ 400	US$ 600	US$ 1000	US$ 2000	US$ 3000
Primary	46.3	36.07	30.4	26.7	21.8	18.6	16.3	9.8
Secondary	13.5	19.6	23.1	25.5	29.0	31.4	33.2	38.9
Tertiary	40.1	44.4	46.5	47.8	49.2	50.0	49.5	48.7

Source: Chenery, Elkington, and Sims. 1970. *Economic Development Report* 148 (July), Cambridge, MA.

[a] US$ in 1964.

value structure as the standard. Some standard structure modes that have a large influence on industrial structure research and take production value structure as the standard can be referred to from Tables 4.1 through 4.4.

4.3.1.2 Standard Structures That Takes the Structures of Labor Force as Indicators In the commodity of society where the labor force flows freely, people tend to engage in the industry with higher income to acquire more money. The labor force distributes in different industries and forms the structure of labor force. The labor force structure is easier to observe compared to production value structure. So the labor force structure rather than the production value structure was investigated at the earliest stages to study the evolution law of industrial structure. William Petty, a British economist from 17th century, pointed out in his book *Political Arithmetic* that manufacturing industry and business get more relative income rather than agriculture. This was because of the fact that most of Holland's population at

Table 4.2 Pattern according to Kuznets

	BASIC LEVEL OF PER CAPITA GDP[a] (%)				
INDUSTRY	US$ 70	US$ 150	US$ 300	US$ 500	US$ 1000
Agriculture	48.4	36.8	26.4	18.7	11.7
Industry and architecture	20.6	26.3	33.0	40.9	48.4
Manufacturing	9.3	13.6	18.2	23.4	29.6
Architecture	4.1	4.2	5.0	6.1	6.6
Commercial and service	31.0	36.9	40.6	40.4	39.9

Source: Kuznets, 1999. *Economic Growth of Nations: Total Output and Production Structure.* Commercial Press, p. 126.

[a] US$ in 1958.

Table 4.3 Pattern according to Chenery, Robinson, and Syrquin

	BASIC LEVEL OF PER CAPITA GDP[a] (%)		
	US$ 140	US$ 560	US$ 2100
Agriculture	29.0	15.7	6.6
Industry and architecture	1.4	1.9	2.1
Manufacturing	23.3	34.6	46.5
Architecture	12.9	15.7	15.3
Commercial and service	33.3	32.2	29.4

Source: Chenery, Robinson, and Syrquin, 1989. *Industrialization and Growth: A Comparative Study.* Shanghai: Shanghai People's Publishing House, pp. 72-75.
[a] US$ in 1970.

that time worked in the manufacturing industry and business, the per capita income of Holland was much higher than that of other nations in Europe. The observation of industrial structure and its advancement by the structure of labor force has still to improve theoretically on some points, but the labor force structure is very clear and statistical data are easy to obtain. Currently, the labor force structure and production value structure are usually used together as indicators for observing industrial structure and measuring its advancement. Some familiar standard structures that take labor force structure as indicators are shown in Tables 4.5 through 4.7.

4.3.2 Relative Comparison and Discretion of the Advancement of Industrial Structure

The path from industrial structure to industrialization differs because of the great difference in the history, resources endorsement, and degree of technology progress in all countries. The "standard structure" is not universal. Therefore, one still needs to use the relative

Table 4.4 Pattern according to Syrquin and Chenery

	BASIC LEVEL OF PER CAPITA GDP[a] (%)					
	LESS THAN US$ 300	US$ 300	US$ 500	US$ 1000	US$ 2000	US$ 4000
Primary industry	46.3	36.0	30.4	26.7	21.8	18.6
Secondary industry	13.5	19.6	23.1	25.5	29	31.4
Tertiary industry	40.1	44.4	46.5	47.8	49.2	50.5

Source: Syrquin and Chenery, 1989. *The World Bank Economic Reviews* 3, pp.145-181.
[a] US$ in 1980.

Table 4.5 Pattern according to Chenery, Elkington, and Sims

	BASIC LEVEL OF PER CAPITA GDP[a] (%)							
INDUSTRY	US$ 100	US$ 200	US$ 300	US$ 400	US$ 600	US$ 1000	US$ 2000	US$ 3000
Primary	68.1	58.7	49.9	43.6	34.8	28.6	23.7	8.3
Secondary	9.6	16.6	20.5	23.4	27.6	30.7	33.2	40.1
Tertiary	22.3	24.7	29.6	33.0	37.6	40.7	43.1	51.6

Source: Chenery, Elkington, and Sims: 1970. *Economic Development Report* 148 (July), Cambridge, MA.
[a] US$ in 1964.

discretion method to carry out the relative analysis. If the height of industrial structure of two countries or two regions is compared, then the relative discretion method will consider one of them as the frame of reference and the other as the one to be analyzed. There are two kinds of comparative methods that can be considered. One method is to compare the similarity between the two industrial structure systems. Their degrees of similarity will be compared to the advancement

Table 4.6 Pattern according to Kuznets

	BASIC LEVEL OF PER CAPITA GDP[a] (%)				
INDUSTRY	US$ 70	US$ 150	US$ 300	US$ 500	US$ 1000
Agriculture	80.3	63.7	46.0	31.4	17.7
Industry and architecture	9.2	17.0	26.9	36.2	45.3
Commercial	4.7	7.2	10.0	12.2	15.2
Service	5.8	12.1	17.1	20.2	21.8

Source: Kuznets, 1999. *Economic Growth of Nations: Total Output and Production Structure.* Commercial Press, p. 126.
[a] US$ in 1958.

Table 4.7 Pattern according to Syrquin and Chenery

	BASIC LEVEL OF PER CAPITA GDP[a] (%)					
	LESS THAN US$ 300	US$ 300	US$ 500	US$ 1000	US$ 2000	US$ 4000
Primary industry	81.0	74.9	65.1	51.7	38.1	24.2
Secondary industry	7.0	9.0	13.2	19.2	25.6	32.6
Tertiary industry	12.0	15.9	21.7	29.1	36.3	43.2

Source: Syrquin and Chenery, 1989. *The World Bank Economic Reviews* 3, pp. 145-181.
[a] US$ in 1980.

of industrial structure. The method is to measure the difference between the two industrial structure systems. The deviation will again be compared to the advancement of industrial structure. We can call the first method "similarity discretion method." The calculation involves the cosine method and the relevant relationship method. The latter method can be called "distance discretion method." The calculation involves the Euclidean distance method and Hamming distance method.

4.3.2.1 Similarity Discretion Method Similarity discretion method takes certain industrial structure as frames of reference and compares the industrial structure system being judged with the system taken for reference by calculating similarity or relative coefficient. This method is to make sure of the degree of the advancement of industrial structure.

4.3.2.1.1 Similarity Coefficient Method This method is deduced from calculation using the cosine method, and is recommended by the United Nations Industrial Development Organization. The formula is given by

$$S_{AB} = \frac{\sum_{K=1}^{n} X_{AK} X_{BK}}{\sqrt{\left(\sum_{K=1}^{n} X_{AK}^2\right)\left(\sum_{K=1}^{n} X_{BK}^2\right)}}, \quad (0 \leq S_{AB} \leq 1)$$

X_{AK} and X_{BK} in this formula represent the weight of sector K in the industrial structure systems A and B, respectively. S_{AB} denotes the similarity coefficient of the industrial structure systems A and B.

4.3.2.1.2 Relative Coefficient Method

$$r_{AB} = \frac{\left(\sum_{K=1}^{n} |X_{AK} - X_A| \times |X_{BK} - X_B|\right)}{\sqrt{\left(\sum_{K=1}^{n} (X_{AK} - X_A)^2\right) \times \left(\sum_{K=1}^{n} (X_{BK} - X_B)^2\right)}}, \quad (-1 \leq r_{AB} \leq 1)$$

In this formula, r_{AB} represents the relative coefficient of the industrial structure systems A and B. X_{AK} and X_{BK} represent the weight of sector K in the industrial structure systems A and B, respectively. X_A and X_B denote the average value of the weight of each sector in industrial structure systems A and B, respectively.

4.3.2.1.3 Distance Discretion Method The key of the distance discretion method is to form a relation formula to calculate the deviation of the two systems. The scholar Yangjun Gong from our country put forward the following formulas for the concrete discretion method of the distance discretion method.

a. Euclidean distance method: The formula is

$$r_{AB} = \left[\sum_{i=1}^{n} \left(X_{Ai} - X_{Bi} \right)^2 \right]^{1/2}$$

b. Hamming distance method: The formula is

$$r_{AB} = \sum_{i=1}^{n} \left| X_{Ai} - X_{Bi} \right|$$

c. Lance distance method: The formula is

$$r_{AB} = \sum_{i=1}^{n} \frac{\left| X_{Ai} - X_{Bi} \right|}{\left(X_{Ai} + X_{Bi} \right)}$$

In order to make the value area of calculation clearer, we revise the above formulas and look for a kind of reflection relation to keep the value area within the interval [0, 1]. Therefore, the above formulas will be changed to the following forms:

a. The new formula for Euclidean distance method is

$$r_{AB} = 1 - C \left[\sum_{i=1}^{n} \left(X_{Ai} - X_{Bi} \right)^2 \right]^{1/2} \tag{4.1}$$

b. The new formula for Hamming distance method is

$$r_{AB} = 1 - C\sum_{i=1}^{n} \left| X_{Ai} - X_{Bi} \right| \tag{4.2}$$

c. The new formula for Lance distance method is

$$r_{AB} = 1 - \sum_{i=1}^{n} \frac{\left| X_{Ai} - X_{Bi} \right|}{\left(X_{Ai} + X_{Bi} \right)}$$

The Equation 4.2 in the above formula is a number which is moderately greater than zero, making r_{AB} fall into the interval [0, 1]. While the choice of Equation 4.2 has much to do with the value of r_{AB}, after being revised, the positive indicators (bigger the value, larger the gap) are changed into inverse indicators (bigger the value, shorter the gap), which are consistent with the commonly used similarity discretion method.

4.4 Economic Development Stage and the Level of Advancement of Industrial Structure

To distinguish the upgrading of industrial structure according to economic development stages, we should find out the stage of industrial structure of the nations or regions, which can be judged by their economic characteristics. The main theories involved in the economic development stage were investigated by W.G. Hoffman, W.W. Rostow, H.B. Chenery and M. Syrquin.

4.4.1 Industry Stage Theory of W.G. Hoffman

The German economist W.G. Hoffman made an explorative research on the evolution laws of industry structure. He pointed out the method in his 1931 publication *The Stage and Type of Industrialization* by making use of time-sequence data of about 20 nations. He put forward the famous "Hoffman laws" after conducting research about the relation between the net production value of consumption material and capital material in the process of heavy industrialization stage.

Table 4.8 Hoffman's Division and Conclusion of the Industrialization Stage

SCOPE OF HOFFMAN COEFFICIENT	STAGE OF INDUSTRIALIZATION
5.0 (±1.0)	First stage
2.5 (±1.0)	Second stage
1.0 (±1.0)	Third stage
Less than 1	Fourth stage

Source: Henderson and Chaloner (translated). 1958. *The Growth of Industrial Economies.* By W.G. Hoffman (Ed.), Manchester: Manchester University Press, pp. 166–170.

The law indicated that in the process of industrialization the ratio of the net production value of consumption material and capital material decreases continuously.

Hoffman also made a division about stages of industrialization. First, Hoffman called the ratio of the net production value of consumption material and capital material as "Hoffman coefficient." And then he divided the process of industrialization into four stages on the basis of changing trends of the Hoffman coefficient. Detailed results are shown in Table 4.8. Hoffman thought that the production of consumption material occupied a dominating position in the manufacturing industry, while the manufacturing of production material was not flourishing in the first stage of industrialization. In the second stage, though the scale of consumption material industry was still much bigger than that of capital material industry, capital material industry started to acquire a faster development rate with regard to its development. In the third stage, capital material industry had a scale equivalent to that of the consumption material industry. In the fourth stage, the scale of capital material industry started to exceed that of the consumption material industry.

The division of the industry stage by Hoffman is for purposes of research. The key point of his research is the heavy industrialization during the process of industrialization. Hoffman's research indicates that the center of industry will shift from light industry (mainly production of consumption material) to the heavy industry (mainly production capital material) with the progress of industrialization.

4.4.2 Economic Growth Stage Theory of Rostow

The American economist Rostow divided the economic growth stage into six stages including the traditional society, the preconditions for take-off, the take-off, drive to maturity, the age of high mass-consumption, and pursuing quality of life according to the technological standard.

Rostow thought that the "take-off" stage is a process for a society to shift from a traditional type to a modern one. The take-off stage of the economy of a country requires some conditions. It should have relative high capital accumulation. Only raising the rate of accumulation and production investment can provide the essential material foundation for economic growth. It needs to build up the dominating sectors that can push the whole economy growth. Dominating sectors drive the development of other sectors by backward, forward, and flanking influence. Backward reviewing is to influence sectors providing production material for it. Forward-looking denotes the instruction function for the emergence of new industry, new techniques, new raw materials, and new energy. Flanking is the influence of the regional society economic development. In accordance with the six economic growth stages, Rostow gave the list of five kinds of dominating sector comprehensive system, shown in Table 4.9.

Rostow thought the sequence of dominating sectors cannot arbitrarily change, and any nation shall have to go through the process of development from a low level to a high level. He also believed that the take-off stage demanded a suitable environment. Because of the fact that the economic system is a subsystem of the whole social development system of a nation, so also "take-off" is a special period in the economic development that needs the corresponding cooperation from the social environment like political system, and so on to make sure of the smooth realization of take-off.

Table 4.9 Division of the Economic Growth Stage according to Rostow

ECONOMIC DEVELOPMENT STAGE	CORRESPONDING PREDOMINANT INDUSTRY
Traditional social	Agriculture as the main body
Preconditions for take-off	Agriculture as the main body
Take-off	Textiles, railways, and architecture
Drive to maturity	Iron and steel, electrical power
Age of high mass consumption	Automobiles
Pursuing quality of life	Service, suburban architecture

Source: Rostow. 1960. *The Stages of Economic Growth: A Non-Communist Manifesto*. Cambridge: Cambridge University Press, pp. 4–16.

4.4.3 *Economic Development Stage Theory of H.B. Chenery and M. Syrquin*

The viewpoint of H.B. Chenery and M. Syrquin was different from that of Rostow. They thought that the economic growth was the overall change in economic structure. In the multinational model set up in the book *Industrialization and Growth: A Comparative Study,* Chenery and Syrquin divided the economic growth into seven stages as shown in Table 4.10.

4.5 Main Influencing Factors of the Advancement of Industrial Structure

The advancement of industrial structure is the result of comprehensive functions of many economic and noneconomic factors. All factors influencing the development of economy directly or indirectly make the function of the industrial structure and promote or restrict the development and change in industrial structure. Overall, the advancement of industrial structure is subject to the comprehensive influence of the national economic development level, resource conditions, science and technology progress, supply of labor force, change in trade, variety of market demand, the change in investment structure, and industrial policies, and so on.

Table 4.10 Division of the Economic Growth Stages according to Chenery, Robinson, and Syrquin

	INCOME LEVEL			
STAGE	US$ IN 1964	US$ IN 1970	STAGE OF INDUSTRIALIZATION	TYPE OF COUNTRIES
0	<100	<140	Production of primary products	Least developed
1	100–200	140–280		
2	200–400	280–560	Initial stage	Quasi-industrialized
3	400–800	560–1120	Middle stage	
4	800–1500	1120–2100	Final stage	
5	1500–2400	2100–3360	Mature stage	Developed and industrialized
6	2400–3600	3360–5040		

Source: Chenery, Robinson, and Syrquin. 1986. *Industrialization and Growth: A Comparative Study.* Oxford: Oxford University Press, p. 71.

4.5.1 Development Level of National Economy

The national economic development and the industrial structure mutually influence each other. Different industrial structure conditions and evolving degrees make the national economy appear to be at different degrees of growth; however, the target and the level of the national economic development during a period not only require the rational change in industrial structure, but also influence industrial structure in various ways, which directly restrain the degree and scope of changes in industrial structure. The quick growth of the national economy can promote the upgrading of the industrial structure, and the decrease of national economic development will restrict the upgrading of the industrial structure. At the same time, the upgrading of industrial structure also increases the development level of the national economy. When the level of the national economic development differs, the ways for the advancement of industrial structure are also different. When the national economic development is at a low level, the only way for the advancement of industrial structure is to exploit the natural resources and make use of a large labor force to produce products with lower added values. When the national economic development level increases, the direction of the advancement of industrial structure will change to technique-intensive industry and knowledge-intensive products with higher added values. The production of low added value products will be transferred to make the economic growth of the regions or nations slow.

4.5.2 Natural Resources Condition

The difference of natural resources endorsement has a great influence on the choice of ways for the advancement of industrial structure. Different industries have different dependence on the natural resources in the three industries. The development of the primary industry is strictly subject to natural resources such as climatic conditions, soil, water, crops, and so on, and the combination of these resources determine the type, and development direction of agriculture. Although the development of the secondary industry is also under the influence of natural resources, the degree of influence is less. However, the industry resources still play a major role in the development of the secondary industry; especially, the type and structure of advantageous resources usually decide the dominating lines and industry structure. The distribution of resources usually decides the industrial layout of the district; the use of resources usually has close connection with investment cost and economic performance of the industry. The development of the tertiary industry basically is independent of the direct influence of nature resources.

Natural resources are the basic elements of economic development and their natural resources endorsement cannot be changed artificially. Although development of economic society depends strongly on the natural resources, this dependence would gradually be weakened along with the progress of technology, development of productivity, improvement of transportation, and so on. For example, with the detection and invention of new energy and new materials, the reliance on the traditional energy and raw materials by the public will become lesser. With the rise of the knowledge-intensive industrial structure, the natural resources endorsement condition tend to have a much weaker influence on the industrial structure to make way for diversified industrial structure development. With technological progress, the abilities to exploit and use resources continuously can be strengthened. And we can improve the supply structure of energy and raw materials and promote industrial structure in the rational direction of development.

4.5.3 Science and Technology Progress

The technological change mainly indicates technical structure change and technological progress, which fundamentally drives the advancement of industrial structure. First, the technical structure change would exert influence on producing technical structure, craft process, rate and mode of production, production scale, market competition and market demand conditions, and provide the valid mechanism for triggering the industry to expand and have a deep impact on the change in industrial structure. Second, the emergence of new technology would also give birth to new industries, reform and eliminate backward industries, and cause industrial structure change. Third, any industry has a corresponding technological level with it. The technological breakthrough of this industry and the extensive application of high-tech methods will result in changes in the structure of this industry and related industries. The breakthrough will cause the development of a series of other related industries through the forward, backward, and flanking connections. Furthermore, it can push the technological changes in related industries with the help of technology proliferation, permeation and induction, and so on. Fourth, difference of technological level determines difference of comparative rate of production among sectors. Among various industries, the dominating industry owning advanced technology absorbs a great deal of innovation results to increase the rate of production and make production factors flow from the sectors with low comparative labor rate to those with high comparative labor rate. Also the innovation results can make production factors flow from the sectors with comparative labor rate increasing slowly to those increasing quickly. And when the dominating industry reaches the mature period, new technological innovations and new dominating industries will emerge because the speed of production rate increases and the cost decreases tend to decelerate. Technological innovation promotes the dominating industry to shift one by one in successive order and has become the apparent characteristic of evolution of industrial structure.

4.5.4 Supply of Labor Force

Being the most important production factor, the amount and quality of labor force supplied have an impact on the perfection and upgrading of industrial structure. For a given amount of labor force supplied, the domestic and international economic development proves that the nation with abundant labor force, cheap labor cost, and lack of funds should develop labor-intensive industry; the nation with abundant funds but lacking in labor force should develop capital-intensive industry. For the quality of labor force supplied, the industry of a higher degree of technology integration should be developed with the supply of high quality labor force. For example, India has the obvious dual economy characteristic. It has a great deal of labor force in the low level and also enjoys a large high quality labor force receiving good education. As a result, the software industry with a higher degree of technology intensive integration develops quickly in India. The quality of labor force in developing country remains at a low level because of the backward economy and low education level. In this situation, depending on the introduction of foreign capital alone cannot quickly promote the economic growth efficiently and the upgrading of industrial structure.

The adjustment of industrial structure is actually the increase of the rate of labor production. The increase of production rate of labor not only indicates the enlargement of production values, but also that much more attention should be paid to the optimization of labor force resources. The change in labor force structure directly influences the change in industrial structure. Despite vast industrialization or the development of information and knowledge industry, the demand for a great deal of new labor force continues. Therefore, how to adjust labor force structure and arrange the right person in the right industry can make the whole labor force structure and industrial structure mutually cooperative. However, continuously adjusting the labor force structure toward the direction of advancement of industrial structure realizes the advancement of industrial structure.

4.5.5 Trade

Trade is an important factor that influences the upgrading of industrial structure. It originates from the difference existing in the industrial structure, which generally denotes the difference of the stages and the industrial structure. Further, it also influences the evolution of industrial structure. It can solve the product problem in the process of upgrading of industrial structure, during which two issues regarding products may arise. One is the case that some products have been produced in excess and the other one is the shortage of products greatly needed in the upgrading process of industrial structure. These two kinds of issues can be solved by trade.

On the one hand, trade can provide a new market opportunity for full play of excessive production ability, which already exists in native districts through the channel of product output and also can stimulate the further upgrading of production ability of native area. On the other hand, the input of products can overcome the bottleneck problems of resources and factors in the native area and create beneficial terms for upgrading of industrial structure. Moreover, the upgrading of industrial structure can be accelerated by the development of trade.

The backward regions should participate in the industry division and absorb mature products and industries transferring from developed regions. Thus, it helps quicken the evolution of industrial structure and stimulate it to upgrade. Both the "flying geese industrial development" theory of Japanese scholar Akamatsu and the "product life cycle" theory of the American scholar Vernon describe this dynamic process.

Trade can be divided into two kinds: domestic and international. The domestic trade develops because of the difference in stages of the industry structure among regions, and the adverse effect works on the upgrading of industrial structure of the regions. International trade emerges because of the difference of state-to-state industrial structure. The formation of international trade also has an important impact on the upgrading of domestic industrial structure.

4.5.6 Structure of Market Demand

The changes in the demand structure and industrial structure have mutual influence on each other. Industrial structure dominates and determines the demand structure, while the demand structure influences and decides the industrial structure too. The change in demand structure causes corresponding change in production and supply structure and therefore corresponding change in industrial structure. The demand structure includes personal consumption structure, the ratio of middle demand and final demand, and so on.

In the demand structure, personal consumption structure has the biggest impact on the change in industrial structure. Personal consumption expectation and structure directly influence the structure and scale of production in the consumption material industry. And this will indirectly influence the development of production material industry required by consumption material industry and then influence the change in industrial structure. With rising income level, not only does the total amount of consumption demand expand, but the consumption structure also changes. Consumption tends to be at a high level and personal demand increases showing multilevel diversification. Multilevel consumption demand structure will cause the progress of multilevel industrial structure.

The ratio between the middle demand and final demand is an important demand structure. The middle demand is a need for the middle product and the final demand is a need for the final product. Middle products are those that still need to go into production process such as raw materials, spare parts, and so on. The final products are those that no longer need to go into production process and can be provided for public consumption or investment.

The demand structure of middle products determines the inner structure of the industry producing middle products. The demand structure of final products determines the inner structure of the industry producing the final products. Naturally, changes in the ratios of middle demand and final demand will cause the industrial structure of social production corresponding changes. The main factors deciding the ratios of middle demand and final demand are as follows. The first one is the professional cooperation level; the higher the professional cooperation degree, the more dependent the final products of the same product on the middle products. The second is the usage rate of production resources. When the usage rate of production resources is higher, the demand for the final product of the same product for consumption of the middle products decreases and vice versa. The third is the function of the final products and complications of manufacturing techniques: the more complicated the manufacturing technique, the more demand for middle products.

4.5.7 Investment Structure

Investment is another important factor of industrial structure evolution. The ratio of investment allocation formed by funds toward different industrial directions is investment structure. Changes in investment structure usually cause changes in the industrial structure. The investment factor has many aspects. From the aspect of the whole, there is the whole social fixed asset investment; from the aspect of various industries, there are basic construction investment and renewal reformation investment; from the aspect of foreign businesses, there are foreign direct investment and indirect investment.

Investment in different directions is the direct reason for changing the existing industrial structure. For the investment to create new demand some new industries will be formed and the original industrial structure will be changed. Some industries will expand at a quicker rate than those that do not receive investment and have an impact on the original industrial structure. The investment for all industries with different investment ratios will cause a difference in the development degrees of various industries and corresponding changes in the industrial structure. As investment is an important factor influencing industrial structure, the government usually adopts investment policies to attain the goal of adjusting industrial structure by adjusting investment structure.

The international investment is also an important factor influencing changes in the industrial structure. Foreign investment will cause the native industry to transfer to the foreign industry. And foreign investment will promote foreign industries to transfer to home industries. These two aspects will cause changes in the domestic industrial structure. Especially, the impact of foreign direct investment on the domestic industrial structure is more direct and profound, which is as follows: The first is that foreign enterprises directly decide what to produce and how much to produce. Changes in the type and quantity of products will directly change the original industrial structure. The second is that changes in the supply structure of middle products and the selling structure of final products from foreign enterprises will directly impact the change in domestic industrial structure. After foreign enterprises enter the domestic market, the supply of middle products can come from both the home industry and the foreign industry. Their final products can also be sold in both at home and abroad. Therefore, the management activities of foreign enterprises cause the change in domestic supply structure and demand structure and promote the change in domestic industrial structure. The third is that the technological innovation of foreign enterprises indirectly impacts the industrial structure of a country or a region.

4.5.8 Industrial Policy

The change in industrial structure is influenced most by the government's industrial policies. Industrial policies are the most important foundations for guiding industry development and industrial structure adjustment. In order to realize the economic development targets established by the government, the government encourages or restricts the development of some industries by setting up industry development strategies and industrial policies. And then the industrial structure will be changed correspondingly. Adjustment of the industrial structure by a government is mainly realized by implementing industrial policies. The government can adjust the various factors including the measures of governmental investment and control, and so on to influence the change in industrial structure. The government can also adjust structures of supply and demand, structure of international trade, and structure of investment by creating public finance policies and monetary policies to influence the change of industrial structure.

5

Relational Analysis for Industrial Output

The industrial structure refers not only to the amount of variety existing in the national economy, the uniqueness, and the nature of existing relationship among industries, but also to quantitative relationships existing in among industries. It refers to the technology and economic relations, which are connected by a variety of industrial inputs and outputs. In essence, the industry relation is an in-depth study on the industrial structure, and the form of relation can be obtained from the notion of "quantitative proportion." It studies the quantitative relationship between "input" and "output." In the input–output analysis, the industrial structure analysis can be used statically to evaluate the current situation and the rationalization. And it also can be used dynamically to analyze the changing trends and factors of the industrial structure.

5.1 Fundamental Tools of the Industrial Structure Relation Analysis

The input–output method is used to analyze the quantitative ratio relationship between the input and output by input–output tables and input–output models, which are fundamental tools for the industrial relational analysis and includes the physical-type table and valuable-type table. The value-based analysis tools are applied in a wide range of fields nowadays.

5.1.1 Input–Output Table

The input–output table is also known as Leontief table or industrial relational table. It uses a matrix to record and reflect the flow of products and services and changing relations in a certain period of time between different departments in an economic system. The input–output table is divided into physical type table and the valuable type table. The former is measured by the substance and the latter by the capital.

5.1.1.1 Physical Input–Output Table The physical input–output table is measured by the standard units or natural units of the product. It is used to show the input and output relationships of the main products in various economic sectors, which are the production and usage relationships as well as the linkages and ratio relationships of the main products between their production and consumption.

In Table 5.1, x_{ij} is the number or quantity of product i consumed or required for producing product j, Y_i the final number of product i, X_i the gross production quantity of product i. L_j is the labor force demand of sector j and L the sum of labor force required for producing a variety of products.

The rows in Table 5.1 depict the distribution and usage of various products and labor force and the distribution and usage of intermediate products and final products. The columns show the number of products and the labor force cost of a variety of products in the production and also show the final products constituted in the community as well as the sum of the variety of products. In the physical type table, the columns of the table cannot be added up and hence does not reflect the dynamical values of the products.

Table 5.1 Simplified General Physical Type Input–Output Table

		OUTPUT								
		INTERMEDIATE PRODUCTS								
INPUT	UNIT	PRODUCT 1	PRODUCT 2	...	PRODUCT j	...	PRODUCT N	TOTAL	FINAL PRODUCT	TOTAL PRODUCTS
Product 1		x_{11}	x_{12}		x_{1j}		x_{1n}	U_1	Y_1	X_1
Product 2		x_{21}	x_{22}		x_{2j}		x_{2n}	U_2	Y_2	X_2
...										
Product i		x_{i1}	x_{i2}		x_{ij}		x_{in}	U_i	Y_i	X_i
...										
Product N		x_{n1}	x_{n1}		x_{nj}		x_{nn}	U_n	Y_n	X_n
Labor force		L_1	L_2		L_j		L_n	L	–	–

The physical input–output table has two balanced equations:

1. The total products are equal to the intermediate products + the final products and is given by

$$\sum_{j=1}^{n} x_{ij} + Y_{ij} = X_i, \quad i = 1, 2, \ldots, n$$

2. The total amount of labor force equals the sum of required amount of labor for the production and is given by

$$L = \sum_{j=1}^{n} L_j$$

The two balanced equations constitute the following:

$$\begin{cases} \sum_{j=1}^{n} L_j = L \\ \sum_{j=1}^{n} x_{ij} + Y_i = X_i \quad (i = 1, 2, \ldots, n) \end{cases}$$

5.1.1.2 Valuable Type Input–Output Table The valuable type input–output table is used widely. The most common one is used to measure a unified monetary unit. Its simplified form is shown in Table 5.2 where x_{ij} refers to the value of products in sector i for producing product j, X_i refers to the year's total product values of sector i, Y_i refers to the year's final product value provided by sector i, and V_i refers to the year's labor salary of sector j.

In Table 5.2, the vertical numbers are the various industrial input structures. These are the numbers of middle products (raw materials) which are purchased by various industries including own industries in order to carry out production and fees paid by various industries for use of production materials, which include wages, interests, and so on. So each column reflects the industrial input structure and the sum is the total investment.

The horizontal data are the output structures in every industry which includes the intermediate products and final products. They

Table 5.2 Valuable Type Input–Output Table

		INPUT										
		INTERMEDIATE PRODUCT					FINAL PRODUCT					
OUTPUT		SECTOR 1	SECTOR 2	… …	SECTOR n	TOTAL	FINAL CONSUMPTION	CAPITAL FORMATION	IMPORT AND EXPORT	TOTAL	TOTAL OUTPUT
Transfer value of production material	Sector 1	x_{11}	x_{12}	…	x_{1n}	x_1	W_1	H_1	F_1	Y_1	X_1
	Sector 2	x_{21}	x_{22}	…	x_{2n}	x_2	W_2	H_2	F_2	Y_2	X_2
	…	…	…	…	…	…	…	…	…	…	…
	Sector n	x_{i1}	x_{i2}	…	x_{in}	X_n	W_n	H_n	F_n	Y_n	X_n
	Total	x_1	x_2	…	x_n						
	Depreciation of fixed assets D	D_1	D_2	…	D_n						
New created value	Workers' compensation V	V_1	V_2	…	V_n						
	Social net income M	M_1	M_2	…	M_n						
	Total N	N_1	N_2	…	N_n						
	Total investment	X_1	X_2	…	X_n						

D_j refers to the depreciation fund of one year of sector j; M_j refers to the society net income of the year of sector j; N_j refers to the national income created by sector j in one year; V_j refers to the labor salary of the year of sector j.

reflect the sales or the distribution of these products. The sum of each row is the total output of the corresponding industrial sector in a certain period of time (such as one year). The column includes the transfer value of production material and newly created values which reflect the value structure of the total products. It constitutes three quadrants. Their economic implications are described next.

1. The first quadrant (the intermediate demand), also known as the endogenous part, is the core of the input–output table. It reflects interdependence and trade relations. Different industries provide the intermediate products to each other in a national social reproduction process during a given period of time (such as one year). So the horizontal array relates to the vertical array of industries. The row data indicate the condition in which a certain industry provides intermediate products to all the industries. The column data indicate the condition that a certain industry purchases intermediate products from all the industries.
2. The second quadrant (final demand proportion) is also known as "exogenous proportion," which reflects the orientation of every industrial product or service in proportion to the final products. The orientation of the final product, that is, the final demand, can be divided into three parts. The first part is the consumption, which can be divided into social consumption and personal consumption. The former refers to the total household consumption, and the latter represents all kinds of social consumption such as public welfare, social security, government administrative expenditures, and so on. The second part is the investment, which is constituted by the update and additional fixed assets; the latter can be divided into productive and nonproductive fixed assets. The third part is the import and export trade.
3. The third quadrant (value added) is also a kind of "exogenous proportion." This part includes two aspects. One part is the retained depreciation of the industrial sectors; the other is the net output (value added) in a certain period of time (such as one year) of the industrial sectors, which is the newly created

value. The net output can be divided into two parts: workers' compensation and net income.

The value type input–output table can be used to establish balanced relations according to the row and column as well as their relationships. The major balanced relations are the following:

1. The balanced relations in the row reflect the flow of products in industrial sectors and show that the semifinished products in the row plus the final products in the row equals the total products in the row; this is given by

$$\sum_{j=1}^{n} x_{ij} + Y_i = X_i \quad (i = 1, 2, \ldots, n)$$

2. The balanced relations in the columns show the output process in forming the value in each industry and reflect the balanced relations between the output in each industry sector and the input in the industrial sectors. Here the transfer value of the product materials in the column plus the newly created value equal the total output value in the column and is given by

$$\sum_{i=1}^{n} x_{ij} + D_j + V_j + M_j = X_j \quad (j = 1, 2, \ldots, n)$$

3. Balanced relationships between rows and columns: first, the total output created by the industry in the row equals the total input in the corresponding column with the same name of the industrial sector. Second, the total amount of the final products equals the sum of the total amount of the national income and the total amount of depreciation of the fixed assets. This is the part of final demand equaling the part of gross added value. Its mathematical expression is

$$\sum_{j=1}^{n} N_j = \sum_{i=1}^{n} Y_i$$

5.1.2 The Input–Output Model

The input–output model is based on the data in the input–output table to calculate various coefficients and then establish the model based on the balance relations of the input–output table.

5.1.2.1 Calculation and Determination of Various Coefficients

5.1.2.1.1 Direct Consumption Coefficient The direct consumption coefficient is also known as the input coefficient, and its economic meaning is the directly consumed quantity of i products when unit j product is produced. The calculation is based on the data of input–output table. Divide the various input components by the total products. The formula for direct consumption coefficient is

$$a_{ij} = \frac{x_{ij}}{X_j} \quad (i, j = 1, 2, \ldots, n)$$

and

$$A = \begin{bmatrix} a_{11} & a_{12} & \cdots & a_{1n} \\ a_{21} & a_{22} & \cdots & a_{2n} \\ \cdots & \cdots & \cdots & \cdots \\ a_{n1} & a_{n2} & \cdots & a_{nn} \end{bmatrix}$$

The matrix A is called direct consumption coefficient matrix. It reflects the technical and economic ties among industries and the technology linkage of products in the input–output table. In the input–output analysis, the direct consumption coefficient is of great significance, and is the most important and fundamental coefficient to establish a model. The larger the value of a_{ij}, the closer the direct economic technology linkage between sectors i and j. However, they are less closer. If $a_{ij} = 0$, then there is no direct technical and economic ties in the two departments. Using the same method we can also calculate other similar coefficients such as the direct depreciation coefficient, payment for labor coefficient, social net income coefficient, and so on.

5.1.2.1.2 Complete Consumption Coefficient The direct consumption of products reflects the relationship between the products in the process of production. However, there is also a complex indirect consumption relationship. This is the quantity of the products or services of one sector consumed by the products or services of another industry sector indirectly. Indirect consumption relationship is multilayered and complicated and reflects the degree of indirect linkage in industrial sectors. Complete consumption coefficient comprehensively reflects the relation between direct and indirect consumption. In other words, the economic meaning of complete consumption coefficient is that the production of one unit output requires consumption of the total products from other sectors directly and indirectly. Compared with the direct consumption coefficient, complete consumption coefficient reflects the economic quantity relations between production in one sector and consumption in other sectors more comprehensively and profoundly. It is important in the correct analysis of the national economy and industrial structure.

Complete consumption coefficient equals the sum of the direct and indirect consumption coefficient and is given by

$$b_{ij} = a_{ij} + \sum_{k=1}^{n} b_{ik} a_{kj} \quad (i, j = 1, 2, \ldots, n)$$

b_{ij} is the complete consumption coefficient; it shows the sum of product i directly and indirectly consumed in producing unit product j. a_{ij} is direct consumption coefficient. $\sum_{k=1}^{n} b_{ik} a_{kj}$ is indirect consumption coefficient, and k stands for the intermediate sector. It shows all the indirect consumption of product i when manufacturing unit product j through k kinds of intermediate goods.

$$B = (I - A)^{-1} - I$$

where matrix B is given by

$$B = \begin{bmatrix} b_{11} & b_{12} & \cdots & b_{1n} \\ a_{21} & b_{22} & \cdots & b_{2n} \\ \vdots & \vdots & & \vdots \\ b_{n1} & b_{n2} & \cdots & b_{nn} \end{bmatrix}$$

$(I-A)^{-1}$ is the inverse matrix of $(I-A)$. The complete consumption coefficient is also known as the inverse matrix coefficient, namely, one unit output change leads to the total changes in other industries directly and indirectly.

5.1.2.2 Two Basic Input–Output Models

5.1.2.2.1 Input–Output Model According to a Row-Balanced Formula By direct consumption coefficient $a_{ij} = x_{ij}/X_j$ we can get $x_{ij} = a_{ij}X_j$, and take it into the row-balanced formula, then we get the input–output model as follows:

$$\begin{cases} a_{11}X_1 + a_{12}X_2 + \cdots + a_{1n}X_n + Y_1 = X_1 \\ a_{21}X_1 + a_{22}X_2 + \cdots + a_{2n}X_n + Y_2 = X_2 \\ \cdots\cdots\cdots\cdots\cdots\cdots \\ a_{n1}X_1 + a_{n2}X_2 + \cdots + a_{nn}X_n + Y_n = X_n \end{cases}$$

The equations of input–output model mentioned above can be converted to $(I-A)X = Y$, where

$$I-A = \begin{bmatrix} 1-a_{11} & -a_{12} & \cdots & -a_{1n} \\ -a_{21} & 1-a_{22} & \cdots & -a_{2n} \\ \cdots & \cdots & \cdots & \cdots \\ -a_{n1} & -a_{n2} & \cdots & 1-a_{nn} \end{bmatrix} X = \begin{bmatrix} X_1 \\ X_2 \\ \vdots \\ X_n \end{bmatrix} Y = \begin{bmatrix} Y_1 \\ Y_2 \\ \vdots \\ Y_n \end{bmatrix}$$

I is a n-factorial unit matrix. $(I-A)$ is called Leontief matrix. Its economic implication is given below. The vertical line in the matrix indicates the relationship between the input and output; every vertical line shows a certain industry investing in the corresponding industrial products for producing a unit product; and the minus symbol indicates the input and the plus symbol indicates the output; and the elements on the diagonal are the net output when products in different industries take out their own consumption. It is clear that the transformation matrix $(I-A)X = Y$ mentioned above shows the relationship between X and Y through the matrix $(I-A)$ and that between the total products and the final products.

5.1.2.2.2 Input–Output Model According to Line-Balanced Formula By the same token, put $x_{ij} = a_{ij} X_j$ into the line-balanced formula and we can get the available input–output model as follows:

$$\begin{cases} a_{11}X_1 + a_{21}X_1 + \cdots + a_{n1}X_1 + N_1 = X_1 \\ a_{12}X_2 + a_{22}X_2 + \cdots + a_{n2}X_2 + N_2 = X_2 \\ \cdots\cdots\cdots\cdots\cdots\cdots \\ a_{1n}X_n + a_{2n}X_n + \cdots + a_{nn}X_n + N_n = X_n \end{cases}$$

The equations of input–output model mentioned above can be converted to $(I - \hat{C})X = N$, in which I is an n-factorial matrix. Matrix X is as the same as mentioned earlier.

$$\hat{C} = \begin{bmatrix} \sum_{i=1}^{n} a_{i1} & 0 & \cdots & 0 \\ 0 & \sum_{i=1}^{n} a_{i2} & \cdots & 0 \\ \cdots & \cdots & \cdots & \cdots \\ 0 & 0 & \cdots & \sum_{i=1}^{n} a_{in} \end{bmatrix} \quad N = \begin{bmatrix} N_1 \\ N_2 \\ \vdots \\ N_n \end{bmatrix}$$

Every element in matrix $\hat{C}$ describes the transfer value coefficient, namely, direct material consumption coefficient; every element in matrix $(I - \hat{C})$ reveals the relationship function between national income and total output value.

In addition to the two most important basic input–output models, according to the requirement, we can also establish the intermediate products flow model, the labor force flow model, and gross national product flow model.

5.2 Analysis of the Relational Effect of the Industrial Structure

The essence of the industry relation is the relationship between supply and demand in the industry. When an industrial production changes, due to the existence of inter-industry linkages of industrial sectors through supply or demand, the change spreads or affects other industries.

5.2.1 *Analysis of Intermediate Demand and Intermediate Input*

The intermediate demand and intermediate input are determined by the input coefficient and is the total of certain industrial outputs that other industries consume for their production. The intermediate input is the total of that which the industries get from other industrial inputs for their production processes.

5.2.1.1 Intermediate Demand Rate The intermediate demand rate in an industry refers to the total demand an industry provides to the national economy divided by the total of the intermediate demand that this industrial product provides to all industries. This is given by

$$G_i = \frac{\sum_{j=1}^{n} x_{ij}}{\sum_{j=1}^{n} x_{ij} + Y_i} \qquad (i = 1, 2, \ldots, n)$$

where G_i refers to the intermediate demand rate of industrial sector i; $\sum_{j=1}^{n} x_{ij}$ refers to the total of the intermediate demand which industrial sector i provides to all industrial sectors; $\sum_{j=1}^{n} x_{ij} + Y_i$ refers to the total output of products in sector i; and Y_i refers to the final demand in sector i. The indicator reflects the number of total products in all industries that are the intermediate products, that is, raw materials for all the industries. The higher the intermediate demand rate, the higher the degree of raw materials industrial sector. The industrial products are either the intermediate products, or the final products; that is to say, "the intermediate demand rate + the final demand rate = 1." So a lower intermediate demand rate in an industrial sector means a higher final demand rate; that is, this industry has the task of also providing the final products. In other words, the products from this industry mostly provide the final demand, or public consumption, or investment consumption or export, and so on. Using the intermediate demand rate, we can accurately calculate the proportion of products of every industrial sector used in the production and consumption and more accurately grasp the status and role of the industrial sector in the national economy.

5.2.1.2 Intermediate Input Rate The intermediate input rate of a certain industrial sector is the proportion of intermediate inputs and the total inputs in a certain period of time (usually one year) in the production process. The rate is given by

$$F_j = \frac{\sum_{i=1}^{n} x_{ij}}{\sum_{i=1}^{n} x_{ij} + D_j + N_j} \quad (j = 1, 2, \ldots, n)$$

where F_j refers to the intermediate inputs rate of sector j; D_j refers to all depreciation fee in sector j (a year); N_j refers to the created value by sector j; The intermediate inputs rate reflects the quantity of raw materials that the industry, in its own production process, needs to purchase from other industrial sectors to produce one production unit, as the industrial total input = the input of intermediate products in the industry + depreciation + net output (value added) = total industrial output value (total output). These are identical equations: the industrial value-added rate = value added/total output value, and value-added rate + intermediate input rate = 1. Therefore, the higher the intermediate input rate in a certain industry, the lower the value-added rate in the industry. The industry with high intermediate input rate is the low-value-added rate industry and vice versa.

5.2.1.3 Effects of the Intermediate Demand Rate and the Intermediate Input Rate The indexes of the intermediate demand rate and the intermediate input rate reflect the industrial relational degree. Its role in the analysis of the industry relation is described next.

First, it can more accurately determine the status of industrial clusters in the national economy according to intermediate demand rates and intermediate input rates. Based on the input–output tables of the United States, Italy, Japan, Norway, and other countries, Chenery,

Watanabe, and other economists divided, after calculation, the industries with different intermediate demand rates and intermediate input rates as shown in Table 5.3.

The four-part industry cluster shown in Table 5.3 forms the main structure with different status and roles in the national economy. That is, parts I, II and III are the material production sectors for the economy of a country, and they provide the intermediate products and final demand products; parts I and II are the intermediate product sectors; the majority of their products serve the final production in part I industry cluster, which has a clear attribute of basic industry; part I industry cluster process the intermediate products from part I and part II, and then put it into the final demand products. Part IV industry cluster, with the exception of the fishing industry, is the mobile intermediate sector for other parts of the industrial cluster.

Further, it can more clearly show the various degrees of interindustry linkages and interdependence. In general, the higher the intermediate demand rate, the greater the output used for raw materials for other industries and more basic the industrial characteristic. Also the higher the intermediate input rate of the industry, the lower the gross added value rate. Using the two-dimensional plan and re-sorting the different industrial sectors in the national economy in the input–output table according to the intermediate input rate and the intermediate demand rate in the industies, that is, considering the horizontal axis from left to right, ranking the intermediate input rate

Table 5.3 Partition of Different Industrial Clusters according to Different Intermediate Demand Rates and Different Intermediate Input Rates

	SMALL INTERMEDIATE DEMAND RATE	LARGE INTERMEDIATE DEMAND RATE
Large intermediate demand rate	(III) Final demand industries: daily groceries, shipbuilding, leather and leather products, food processing, transport equipment, machinery, timber, wood processing, nonmetallic mineral products, other manufacturing goods	(II) Intermediate goods industry: steel, paper and paper products, petroleum products, nonferrous metal smelting, chemical, coal processing, rubber, textile, printing and publishing
Small intermediate demand rate	(IV) Industries of final demand: fisheries, transportation, commerce, service industry	(I) Industries of middle products: agriculture, forestry, coal, metal mining, oil and natural gas, nonmetallic mining, power

Source: Yang Zhi, *Introduction of Industry Economics*. Beijing, China: People's University Press, 1985, p. 105.

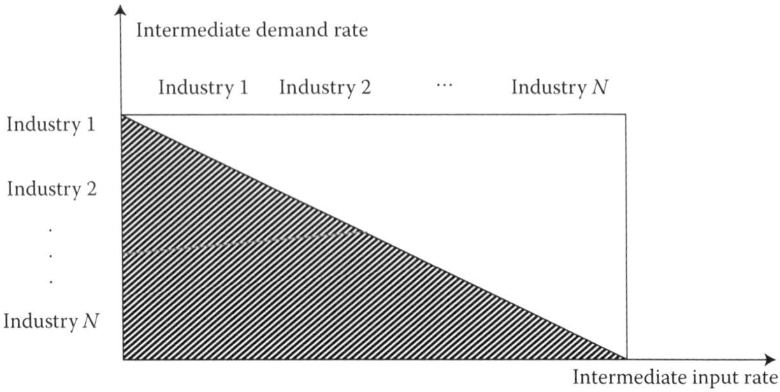

Figure 5.1 Triangular graph of input coefficients.

from big to small, and the vertical axis from top to bottom, ranking the intermediate demand rate from small to big, we obtain the rearranged input coefficient table as shown in Figure 5.1.

In the triangular graph of input coefficients, all input coefficients appear only in the lower triangle. This means that the x_{ij}'s, which adjust the order in the input–output table, the data present only in the shadow of the lower triangle, and x_{ij}'s in the upper triangle are all zero. It means that the intermediate demand rate of industry 1 is zero, and its intermediate input rate is 1. Industry 1 has no intermediate demand, and all products are the final products. And it will get all intermediate inputs from industries 2, 3, ..., n. Industry 2 is needed only by industry 1, and other industries do not have intermediate demands from industry 2. Most products in industry 2 are the final products and, at the same time, industry 2 gets the intermediate input from all other industries except for industry 1, and so on up to industry n. The intermediate demand rate of industry n is 1, and its intermediate input rate is zero. This means that the products in industry n are all intermediate products, and at the same time, industry n does not need to get any intermediate inputs from other industries.

Overall, as seen from the bottom of the lower triangle, industry n need not obtain intermediate inputs from other industries in the production process, and its output, which is the intermediate demand, entirely provides other industries products for consumption, thus not providing the final products to society. Such an industry is a basic industry of the whole industrial structure. On the other hand, at the top of the triangle, all the outputs are the final products which are provided to the community for final consumption, and in their production process need to absorb the output of other industries as part of their intermediate inputs. These industries have the main task of meeting the final demands of the community and, therefore, can be regarded as the final product industries. Under normal circumstances, the intermediate input rate of the final industry is larger.

In addition, it can reveal the order of coordinated development in various sectors. The so-called "order" in the industrial development refers to the significance of the contribution of the industry to the operation of the national economy. This order is determined by the degree of interdependence in the industrial relation. From the triangle configuration, it is not difficult to understand why the industries must maintain a certain sequence, and must pay attention to the degree of affinity and interdependence among industries. Because basic industries provide raw materials for other industries for the existence and development of other industries, they must be developed first. That is why in the development of the national economy priority is given to develop basic industries such as agriculture, energy, and so on.

5.2.2 *Breadth and Depth of Industry-Related Effects*

The degree of inter-industry linkage can be examined in two ways: by studying the linkage breadth and the linkage depth between the industries.

The breadth of inter-industry linkages can be measured and studied by the direct consumption coefficient a_{ij}. When $a_{ij} = 0$, sectors i and j have no direct relationship; when $a_{ij} > 0$, and with more sectors j $(j = 1, 2, \ldots, n)$ involved, the industrial sector j has broader relationship with other industries and vice versa. It should be noted that, when $a_{ij} = 0$, it is not possible to conclude that the sectors i and j have no complete linkages. In fact, any two or more industrial sectors can have no direct relations. However, whether they have complete relation, depending on the complete consumption coefficient b_{ij}, indirect consumption is inclusive.

The relational depth of industries can be measured by calculating the proportion which the respective flow in the lists of input–output tables (that is, the industry's direct consumption) takes in the total dircct consumption. The measurement index can be calculated by the following formula:

$$r_{ij} = \frac{x_{ij}}{\sum_{i=1}^{n} x_{ij}} \quad (i, j = 1, 2, \ldots, n)$$

In the formula, the denominator indicates the quantity of total direct consumption of products of other industrial sectors in the production process, which sector j consumes. The numerator indicates the quantity of direct consumption of products of the specific sector i, which sector j consumes. r_{ij} is a ratio of the two, also called the linkage depth for sector j to sector i. The greater r_{ij} value shows that the sector j consumes more products of sector i in the production process, and that sector j has feeper links with sector i; vice versa, the linkage depth is shallower. Calculating all r_{ij} values can reflect the degree of depth of the relation between the industries.

5.3 Inter-Industry Effect Results Analysis

In the system of national economy, industry effect refers to the phenomena that occurs when a particular industry sector changes and the change is passed *quo modo* along the different industrial sectors to which it is directly related. The industrial process is affected, although the influence wanes. The affected industrial system impacts the national economy.

The effect results analysis in the industry is to analyze changes and effects when certain industrial changes lead to development and change in other industries. These changes and effects are mainly the results of some of the data causing other data changes in the input–output tables.

5.3.1 Industrial Sensitivity Coefficient and Influence Coefficient

The production activities of any industry sector are bound to be affected or influenced by the production of other industrial activities because of the relation that exist between the industries. Here, the impact level of an industry on other industries is called the force of influence; the impact level on an industry by other industries is called the degree of sensitivity. Different industries have different forces of influence and degrees of sensitivity. These are obtained by using the influence coefficient and sensitivity coefficient, respectively.

5.3.1.1 Influence Coefficient The influence coefficient formula is given by

$$\gamma_j = \frac{\sum_{i=1}^{n} b_{ij}}{\frac{1}{n}\sum_{i=1}^{n}\sum_{j=1}^{n} b_{ij}} \quad (i, j = 1, 2, \ldots, n)$$

where γ_j is the influence coefficient of sector j impacting on other industry sectors and b_{ij} refers to the coefficient in the complete consumption coefficient matrix B. If the industry influence coefficient is greater than 1, the industrial influence force is above the average level of the all industries ranking; if industry influence coefficient is less than 1, the industrial influence force is below the average level of the all industries ranking.

5.3.1.2 Sensitivity Coefficient The sensitivity coefficient formula is given by

$$\mu_i = \frac{\sum_{j=1}^{n} b_{ij}}{\frac{1}{n}\sum_{i=1}^{n}\sum_{j=1}^{n} b_{ij}} \quad (i, j = 1, 2, \ldots, n)$$

If the sensitivity coefficient is greater than 1, the sensitivity degree is above the average level of the all industries ranking; if the sensitivity coefficient is less than 1, the industrial influence force is below the average level of the all industries ranking.

5.3.2 Industrial Production-Induced Coefficient and Final Demand-Dependent Coefficient

5.3.2.1 Production-Induced Coefficient Production-induced factor is used to calculate the degree to which the final demand items of each industry sector (such as consumption, investment, exports, etc.) induce production. The appropriate production-induced factor can be derived through input–output table, and reveals the degree of effect which the final demand product has on the various industrial sectors.

The so-called industrial production-induced coefficient is the total of the corresponding final demand products divided by the industrial production-induced quantum in a variety of final demand products.

$$W_{il} = \frac{Z_{il}}{Y_l} \quad (i, l = 1, 2, \ldots, n)$$

Among them, W_{il} is the production-induced factor of the final demand product l in sector i; Z_{il} is the production-induced quantum which sector i induces on the final demand product l; Y_l is the total of final demand quantum which all industries induce on the final demand product l.

5.3.2.2 Industrial Final Demand-Dependent Coefficient The industrial final demand-dependent degree refers to the degree of dependence which a certain industrial production has on each final demand products (consumption, investment, exports, etc.). It includes both the direct and indirect dependence. It is calculated by dividing the sum of production-induced quantum of the final demand product in the industry by the production-induced quantum of the final demand product in the industry. It is the degree of dependence Q_{il} that this industry has on the final demand product.

$$Q_{il} = \frac{Z_{il}}{\sum_{L=1}^{n} Z_{il}} \quad (i, l = 1, 2, \ldots, n)$$

Among them, Q_{il} is the degree of dependence of sector i on the final demand product l; Z_{il} is the production induced-quantum of final demand product in sector i.

By calculating the degree of dependence of each industrial production on the final demand product one can get the final dependence coefficient table. Analyzing and classifying this table one can find that some production sectors do not have direct relationship with consumption; however, eventually their large output depends on consumption through indirect relationship. On the other hand, that each industrial production eventually depends on the investment, consumption, or export is made clear. Accordingly, the industry can be classified as consumption-dependent industries, investment-dependent industries and export-dependent industries, and so on.

5.3.2.3 Comprehensive Employment and Capital Coefficient The inverse matrix $(I - A)^{-1}$ calculates not only the change in the final output quantum induced by the final demand change, but also the number of employees for the final demand and the amount of capital with each growth in industrial production. This will require calculating the comprehensive employment coefficient and integrated capital coefficient.

5.3.2.3.1 Comprehensive Employment Coefficient The economic meaning of comprehensive employment coefficient is that industries directly and indirectly need employment population for unit production. It is clear that different industries have different comprehensive employment coefficients.

The comprehensive employment coefficient in industry i = employment coefficient in industry $i \times$ the coefficient in industry i listed in the completely consumed matrix.

The employment coefficient is the employment population per unit output.

$$\text{Employment coefficient in sector } i = \frac{\text{Employment population in sector } i}{\text{Total output in sector } i}$$

5.3.2.3.2 Comprehensive Capital Coefficient The integrated capital coefficient refers to amount of capital needed by industrial sectors directly and indirectly in order to produce one unit. The formula is given by: comprehensive capital coefficient in industry i = the capital coefficient in industry $i \times$ the coefficient in vertical line i in the completely consumed coefficient matrix.

$$\text{Capital coefficient in industry } i = \frac{\text{Capital amount in industry } i}{\text{Output in industry } i}$$

In general, the investment in public industries and the basic industries such as transport, posts and telecommunications, gas supply, and so on reflects a large capital coefficient. In the manufacturing sector, the capital coefficient is higher in industries such as cement, iron and steel, chemicals, and so on.

Attention should be paid to the following problems while doing effect results analysis. First, the stability and effectiveness of the coefficient should be considered. As the input–output table is based only on a past era of inter-industry production technique linkages, the input coefficient reflects only the inter-industry input situation at that time. With the development of national economy and improvement of production technology, the input coefficient is bound to change along with scale change in the economy. Such a change is not large in the short term, but will be large in the long term. In this way, given the past input coefficient, we can analyze the future short-term industrial effect results. But for future long-term analysis of the industrial effect results and for ensuring the accuracy of the analysis we must amend and predict the available input–output coefficient and maintain its stability and effectiveness. When amending, it is necessary to pay special attention to the technological change in certain key sectors as well as the impact of technology on the input coefficient.

Second, the phenomenon of time lag that influences the results should be considered. The time lag on the industry influence results means that a final demand change in a certain industry leads to other industrial changes and does not immediately reflect in the output changes. In other words, there is a time process for a final demand change in a certain industry leading to other industrial changes. The length of the time process often varies depending on the industries, stages of economic cycles such as prosperity and depression. Such differences are often due to the existence of "stock." When the demand increases, it first reflects in the stock reduction. In this way, the effect led by the final demand changes remain suspended or reduced due to the existence of stock. On the other hand, when the stocks are insufficient to meet the increasing demand of the influence results, and the production cannot immediately increase, the influence results led by the demand changes may be reflected in the price improvement. It is clear that the role of buffer stock reflects in the stock list of the final demand in the input–output table; but the intermediate demand and the intermediate input matrix cannot reflect this fluctuation in economy. As a result, when we analyze the influence results of the industries, we have to consider the time-delay phenomenon so as not to draw erroneous conclusions.

6
Regional Industrial Structure

6.1 Conception and Characteristic of Region

6.1.1 Regional Classification

Region usually refers to the sequential economic zone in the earth's surface, and sometimes it also refers to the geographical area or administrative region. Region is the form of social production and geographical division, and it objectively exists like the social production and the geographical division. In the formation and development of a region, according to the different principles of division, the region can be divided into three: homogeneous region, the nodes region, and planning region.*

6.1.1.1 Homogeneous Region This is a broad division method, which makes different regions within the same economic zone have similar conditions and development levels in natural, economic, and society. The main purpose is to deal with the relationship among different regions in different development levels. For example, the Chinese "The Ninth Five-Year Plan of Economic and Social Development Long-Term Goals for 2010" divides China into seven multiprovince-city economic regions. They are the Yangtze River Delta and the nearby regions, the Bohai Sea region, the southeast coastal region, parts of Chinese Provinces in the southwest and south, the north-east region, five provinces in the central region and the northwest region of China. The partition of economic zone eliminates imbalance factors of development within the region, and helps the country as a whole to maintain the basic directions of economic development

* Deng Weigen. *Industry Economic Research*. Beijing: Economic and Management Publishing House, 2001.

and is convenient to deal with the relationship between the developed regions and the less developed regions of the country.

6.1.1.2 Nodes Region The nodes region is also known as the polarizing region. It is the specialized economy region, which compartmentalizes the city center and transportation hub with the national significance. It is made up of interrelated regions, which cincture the city center and have specialized functions. Large economic regions mainly function on the inherent links of economic activities and natural geography characteristics to compartmentalize the polarizing region. This division puts economic benefits as the core and emphasizes the production links and the rationalization of the industrial structure. The division aims to play the important role of the economic center in promoting the region's economic development. And this division method will also help the economy center play a shining role.

6.1.1.3 Planning Region Planning region is also known as the administrative region. This region is compartmentalized according to the coverage of the executive power. This division provides a clear boundary and is convenient to implement economic management measures for the government. In China, administrative division economic zone includes province–city-level economic zones, region–city-level economic zones, and county–city-level economic zones and so on. The province–city-level economic zones with national significance are important and indispensable for a large country to carry out macroeconomic management, and they are often consistent with the corresponding polarizing region and have strong practical significance.

6.1.2 Characteristics of Regions

The concept of a region lies between the nation and enterprises. Compared with the country, regions have the following important characteristics.

1. *Greater openness.* Regions cannot be closed systems. They not only export their specialized products to the outside regions, but also import the necessary products from the outside. And

they also have to keep relatively stable technological and economic ties with other regions.

2. *Nonequilibrium more prominent.* The reason why regions exist in relative independence is because regions have unique local characteristics that are different from that of the external environment. This local characteristic originates from the inter-region differences of production resources endowment and focuses on the regional advantages.
3. *Indivisibility in economics.* From points of view of trade, nations often have a multiplicity of man-made obstacles, but regions do not.
4. *Limitations on power.* Regional autonomy is less than that of a nation.
5. *Duality.* On the one hand, regions should be special areas that are integrated in the economy as far as possible. On the other hand, because of the characteristics of nature, cultural accumulation in history, the residents and their contribution to the production capacity, they are a tache in the total chain of links in the national economy.

6.2 Optimization of Regional Industrial Structure and Economic Development

6.2.1 Meaning of Regional Industrial Structure

Regional economy is a component of the national economy. It is built on the regional resources advantage. So, generally the regional economy does not require to include all industries. Most of the regions have industries only partially interrelated with their resources advantage. However, the region is not supported by a single industry. A number of industries jointly promote the regional economic development. Then the relationship formed between different industries in the same region and inter-region is the regional industrial structure. The so-called regional industrial structure is the industrial composition state, development level, production links, and the quantitative ratio relationship among industries of the region. It includes two aspects. One refers to proportion relationships on the scale among industries. It is directly related to the balance of structure; the other refers to the links and conjunction modes among industries. It is directly related to the structure advancement and the effectiveness. The rate inter-industry relationship constitutes the quantitative aspect of regional industrial structure and inter-industry linking ways constitute its qualitative aspect. The regional industrial structure is the unification of quality and quantity, and both aspects are indispensable.

In the final analysis, the result is the area of human economic activities and space division in the long-term decided by the internal mechanism of regional economic development process. The various economic benefits obtained from the labor division are the direct power and underlying causes of the birth and development of regional industrial structure. Various natural conditions of the region and endowment of different elements form the comparative advantages of the region and effect a regional labor division. So every industry in different regions has different distribution conditions. At the individual area level, the regional labor division combines various production taches of social reproduction and various production department to make specialized regional centers; In the national economic level, regional labor division puts up the division labor collaboration among regions. Therefore, the industrial structure of the region, which is a part of the system of the geographical labor division of the national economy, is essentially a combination of economic structure and spatial structure with effective allocation of resources. The reasonable requirement of regional industrial structure is to make the specialization as the core and organize tens of thousands of businesses to link by production, distribution, or technology. The proportion of inter-sector coordination is complementary to maximize economic benefits. The regional industrial structure is very complicated, which includes the relationship between the national industrial structure and the regional industrial structure, between different regional industrial structures, and the relationship within the regional industrial structure. It is not only a major part of the regional economic structure, but also the subsystem in national industrial structure. Regional industrial structure includes not only the sector structure, but also the spatial structure. The sector structure refers to the proportion among the primary, secondary, and tertiary industries; and the space structure refers to the choice and optimization in a certain region among the leading industries, supporting industries, matching industries, basic industries, and strategic industries. The industrial sector structure reflects the timing in the development of industrial structure; and the spatial structure mainly reflects the industrial spatial combination. They achieve the reunification in the same region. The regional industrial structure should be established on a reasonable geographic division of labor; and it is not only an indispensable component in the nation geographical division of labor, but can also be further divided into relatively integrated low-level structures, and then form the multilevel labor division system. In each region the industrial structure is not only an integrated part in the system, but also has the relative independence and integrity.

6.2.2 Characteristics of Regional Industrial Structure

According to conceptions of the industry economics and characteristics of the regional economy, regional industrial structure has the following main features:

1. *Strong nonequilibrium:* The regional economy is not a completely independent economic complex. Then its industrial structure should not blindly pursue the "large and full." The region should develop the homologous advantage industries according to the resource endowment in the area and then form the nonequilibrium of regional industrial structure.
2. *A certain degree of comprehensiveness:* On the basis of characteristics of the nonequilibrium, the regional industrial structure also has certain comprehensiveness. The region is formed by a variety of combination and development of production factors on a certain space. It is a miniature of the national economy. It requires each region to develop three industries as far as possible. It also requires the region should have more specific infrastructure to safeguard the normal operation of the regional economy and support the growth and development of the leading industries.
3. *Higher degree of openness:* The regional industrial structure is an open system. As a major part of the regional economic structure, it is closely linked with other parts of the economy. The formation and evolution of the regional industrial structure are both in the large environment of regional

economic exchanges and regional exchanges and they also have a wide range of the exchanges with the environment. In today's research about regional industrial structure, it must be placed on the national economy background as well as the globalization background to consider its rationalization and advancement. This is because the enterprises in the market economy must be open and able to flow so that it can survive and develop in the global market.

4. *Systemic change:* The interior of the regional industrial structure has complex interaction. Regional industrial structure is an organic whole. Within the economy region, different industries and different trades in the industry are mutually linking and constraining. The regional industrial structure has positive feedback effects on all aspects and at various levels, which make the system unstable. And there is also the saturation effect that restricts the economic growth, which is called the negative feedback. When changes of the regional industrial structure state exceed the threshold, the original regional industrial structure becomes unstable and will gradually give way to form a new industrial structure.

6.2.3 Optimizing Regional Industrial Structure and Economic Development

The regional industrial structure optimization is a dynamic process. At different stages of development and conditions the regional industrial structure optimization has different connotations. In general, it includes rationalization and efficiency of the industrial structure. And in the current era, it also includes the internationalization of industrial development, the improvement of the industrial quality and competitiveness.*

The regional industrial structure rationalization mainly enhances the ability to coordinate and raises the level of association between industries in the system of regional economy. It focuses mainly on four aspects. First is the coordination between industry qualities; it does not have technical level deficiency between industries, and there

* He Weida. *WTO and Chinese Industry Upgrade.* Beijing: China Audit Publishing House, 2000.

is no strong disparity in labor productivity. Second is the coordination in the relative position between industries. At a certain stage of development, various industries are in different positions because of different growth rates; they form orderly permutation and combination of different degrees. The third is the coordination links between industries. Relations of interdependence, mutual services, and mutual promotion occur according to the links in production technology. The fourth is to maximize the structure effectiveness. The dynamic and well-balanced ratio between inputs and intermediate products, intermediate products and final products, final products and consumers relationship is maintained. On the one hand, industrial structure rationalization is made; and on the other, the industrial structure should have the ability to achieve further advancement at the same time. The industries can take full advantage of the benefits of the international division and national resources at a certain stage of economic development within the conditions of the industrial structure rationalization; the resources can be allocated reasonably and used effectively. The industries can develop in a coordinated fashion and form a constantly evolving mechanism for the industrial structure upgrading; the social production, exchange, distribution, consumption, and expansion of reproduction can be carried out smoothly. The virtuous cycle for industrial structure development is formed.

Reasonable regional industrial structure should be in accordance with the following criteria: (1) full and effective use of the region's human, material, financial, and natural resources; (2) suitability for inter-region division and the international division of labor forces; (3) coordination of different sectors in the regional economy and smooth running of the economy and societal-expanded reproduction; (4) growth of the regional economy at a steady pace and adaption to changes in market demand; (5) harmonious development of industries and economics; (6) harmonious development of labor force, resources, capital, technology, and knowledge-intensive industries; and (7) realization of sustainable development among population, resources, and the environment.

Reasonable and coordinated regional industrial structure is an important prerequisite for the regional economic growth. On the one hand, when the regional industrial structure is unreasonable or lacks coordination, it is essential for the regional industrial structure to adjust and optimize; on the other hand, the rationalization and advancement of the regional industrial structure requires promotion of sustainable and rapid regional economic development. Therefore, optimization of the industrial structure will promote rapid economic development, thereby promoting regional industrial structure optimization.

The advancement of regional industrial structure lies in its upgradation and evolvement, which includes the choice of leading industries, cultivation of strategic industries and the recession of sunset industry. It is a dynamic process going from a low level to a high level. The main expression is the subrogation of leading industries order. In general, it is the subrogation for three industries evolving from the primary industry to the secondary industry and the tertiary industry. It also transforms from the labor-intensive industries to capital-intensive industries and technology-intensive industries. In the same industry, it evolves from manufacturing more primary products to manufacture more intermediate and the final products. It evolves from the rough and shallow-processing to elaborate-processing and deep-processing industries. It evolves from low value-added to high value-added industries.

The regional industrial structure is changing along with the economic growth, which constitutes the regional economic development and national economic development. The long-term growth of regional economy is highly sensitive to the changing industrial structure and is one of the key factors affecting the quality and the improvement of regional economic growth. The strength and weakness of regional industrial structure are important symbols of the quality and level of regional economic development. The regional industrial structure evolution determines the regional industrialization and modernization process. The regional industrial structure is a subsystem of the overall national industrial structure, and its development profoundly impacts on the national economy. The development of the national economy is impacted not only by various regional industrial structures, but also by all relationships among the regional industrial structure. To achieve high effective growth and development of the national economy, we have to coordinate relationships among regional industrial structures better. Therefore, optimization of the regional industrial structure is one of the essential conditions which promote the regional economy and even the entire national economy for sustainable, rapid, and healthy development. China has a huge geographical territory and the economic development level is extremely imbalanced. So it is particularly important that regional industrial structure is optimized.

6.3 Influence Factors of Regional Industrial Structure

The major factors which impact the regional industrial structure are some special regional conditions.

6.3.1 Regional Factor Endowment

The specificity in the regional production factors differentiates different regional industrial structures. The region with rich labor force is conducive to develop the labor-intensive industries and form the regional industrial structure dominated by the labor-intensive industries. The region with technically skilled and educated labor force is conducive to develop technology-intensive or knowledge-intensive industries and form the regional industrial structure dominated by technology or knowledge-intensive industries. The region with rich mineral resources is conducive to develop the resource-based industries and form the regional industrial structure dominated by resource industries. The region with adequate capital is more conducive to form the regional industry structure dominated by capital-intensive industries. The regional production factors such as labor force, capital, technology and resources are decisive factors in the regional industrial structure. Regional economic development must be based on regional factor endowment conditions.

The real level of the regional industrial structure cannot be divorced from the supply conditions of regional production factors. And the optimization of the regional industrial structure must be from the factual supply conditions of regional production factors.

6.3.2 Demand-Oriented Structure

The demand structure is the driving force behind the evolution of industrial structure. The strong consumer demand provides a broad market for the regional industrial development and provides a market assurance for industrial expansion. Changes in the demand structure will cause changes in industrial structure. Enhancement of regional consumption levels will lead to upgrading of the consumption structure so as to promote the advancement of industrial structure. At the same time, under the conditions of inter-regional cooperation and labor force division, the regional industrial structure is also affected by other regional demand structure, especially in the region of higher consumption. The international labor force division affects the regional industrial structure by the international market structure especially in the higher consumption-level countries. The higher the openness of the economy in a region, the higher the interrelation and interaction degrees between the regional industrial structure and demand structure in other regions of the world market.

6.3.3 Inter-Regional Economic Relation

Inter-regional economic links are inter-regional commodity trade and the production factors flow between regions. The more close the inter-regional economic links, the more effectively the industry structure plays its role. Inter-regional trade is the only way to achieve the regional comparative revenues. A division of labor force among regional industrial structures is formed by comparing the economic advantages in different regions. Inter-regional flow among production factors will improve the supply situation of the region and optimize the regional industrial structure. In an open economy, the regional industrial structure is also impacted by external and foreign markets. International goods trade and elements movement put into effect in a certain region in one country directly impact the industrial structure of the region and indirectly impact other regional industrial structure through the inter-regional links.

6.3.4 Regional Concentration Degree of Production

The regional concentration degree of production refers to the production scale proportion which the industrial products of a country in a certain region occupying a proportion in the whole country. It can be expressed by the location quotient. The regional concentration degree of production is influenced by the industrial distribution strategy, business environment, market size, and other production factors of the country.

Different regional concentration degrees of production in the different industries determine the different regional industrial structure.

In addition, the political and law and order environment of the region, economic policy environment, traffic conditions, historical and cultural backgrounds, and so on will also affect regional industrial structure in different degrees. These factors together determine the current status and the direction of development of the regional industrial structure as they are the foundation and approach for the optimization of the regional industrial structure.

6.4 Economic Indicators to Analyze the Regional Industrial Structure

6.4.1 Economic Indicators Reflecting the Specialization of Regional Industrial Structure

Specialization of regional industrial structure refers to the concentration degree in one or more industries in a region related to the national level. The level of specialization can be measured by the production value, output, labor force, amount of fixed assets, and so on. The following indicators can be used to reflect the specialization of regional industrial structure.

6.4.1.1 Location Quotient Location quotient (LQ) is also known as the concentration degree of production region. It can be used to indirectly reflect the structure and direction of inter-regional economic links by relative specialization degrees of different industry sectors in different regions. The common indicators are output values, product quantity, employment number, fixed assets, and so on. The formula is given by

$$LQ = \frac{\text{The number of employees in sector A in a region/} \atop \text{The total number of employees in the region}}{\text{The number of employees in sector A in the nation/} \atop \text{The total number of employees in the nation}}$$

$$= \frac{\text{The number of employees in sector A in a region/} \atop \text{The number of employees in sector A in the nation}}{\text{The total number of employees in the region/} \atop \text{The total number of employees in the nation}}$$

The number of employees in the above formula can be substituted by output values, product quantity, and fixed assets, respectively, to calculate each location quotient. Using the number of employees, output values, product quantity, and fixed assets to calculate the location quotient are called regional labor force concentration degree, product concentration degree, output concentration degree and fixed assets concentration degree, respectively.

If $LQ > 1$, the degree of professionalization in industry A in the region exceeds that of the national level and belongs to specialized departments; if $LQ < 1$ the degree of professionalization in industry A in the region is less than that of the national level and necessitates import of products from outside of the region; if $LQ = 1$ shows that the degree of professionalization in industry A in the region is equal to the national level and there is basic self-sufficiency. The greater the LQ value, the higher the level of professionalism.

6.4.1.2 Per Capita Output Coefficient and Per Capita Production Value Coefficient These two coefficients reflect the labor productivity rate in the region compared to the same industry in the nation. The formula is given by

$$\text{Per capita output coefficient} = \frac{\text{Per capita output in industry A in a region}}{\text{Per capita output in industry A in the nation}}$$

$$\text{Per capita production value coefficient} = \frac{\text{Per capita production value in industry A in a region}}{\text{Per capita production value in industry A in the nation}}$$

If the per capita output coefficient and per capita production value coefficient in a region are greater than 1, it shows that the labor productivity rate in the region is higher than that of the national average value. At the same time, it is always the specialized sector in the region.

6.4.1.3 Regional and Inter-Regional Commodity Rate The indicators of the inter-regional export of the regional industries. In an open regional economic condition, the products are bound to be exported to other places at home and abroad. Hence, there exists inter-regional export of the regional industries. This indicator also reflects the professional level of the industrial region. The main indicators are as follows:

$$\text{Regional commodity rate} = \frac{\text{Product output in industry A in the region}}{\text{Capacity in industry A in the region}}$$

$$\text{Interregional commodity rate} = \frac{\text{Product output in industry A in a region}}{\text{Product output in industry A in the nation}}$$

In general, the regional commodity rate and inter-regional commodity rate in a regional specialization industry are ahead of other regions in the country. The higher regional commodity rate shows that there are more goods for export and the level of professionalism in the region is higher. The higher the inter-regional commodity rate shows that this regional industry export is more important in the nation. So this region has a higher degree of specialization.

6.4.1.4 Coefficient of Regional Industrial Output This indicator is sometimes used to describe the specialization degree of the regional industry, which involves a number of industries. So it is always calculated by the ouput value. The formula is as follows:

$$\text{Coefficient of regional industrial output} = \frac{\text{Output of industry A in a region}}{\text{Output of every industry in the region}}$$

6.4.2 Economic Indicators Reflecting the Convergence of Regional Industrial Structure

The industrial structure between different regions may be very different or similar. The greater this difference, the higher the division level. This also indicates that the complementarity of the regional economies are much stronger and the inter-regional links are more closer. On the contrary, the more similar the industrial structure in the regions, the lower the division level. This also indicates that the complementarity of the regional economies is much weaker and the inter-regional links are less close. The convergence or the degree of difference in the regional industrial structure can be scaled by the correlation coefficient and similar coefficients which are recommended by the United Nations.

Correlation coefficient:

$$r_{ij} = \frac{\sum_{k=1}^{n}(X_{ik} - \bar{X}_i)(X_{jk} - \bar{X}_j)}{\sqrt{\sum_{k=1}^{n}(X_{ik} - \bar{X}_i)^2 \sum_{j=1}^{n}(X_{jk} - \bar{X}_j)^2}} \quad (-1 \le r_{ij} \le 1)$$

where r_{ij} represents correlation coefficient in regions i and j; X_{ik}, X_{jk} represents the proportion of sector k takes in regions i and j of regional industrial structure; $\bar{X}_i$, $\bar{X}_j$ represents the average proportion value of regions i and j take in industrial structure in their sectors; $r_{ij} = -1$ indicates that the industrial structure of regions i and j are completely opposite; $r_{ij} = 1$ indicates that the industrial structure of regions i and j are completely identical; the closer the value of r_{ij} to 1, the higher the industrial structure convergence degree in the two regions.

Similarity coefficient:

$$S_{ij} = \frac{\sum_{k=1}^{n} X_{ik} X_{jk}}{\sqrt{\sum_{k=1}^{n} X_{ik}^2 \sum_{k=1}^{n} X_{jk}^2}} \quad (0 \le S_{ij} \le 1)$$

where X_{ik} and X_{jk} refer to the proportions sector k takes in regions i and j of regional industrial structure; S_{ij} refers to the similarity coefficient in regions i and j of the regional industrial structure. S_{ij} is in the interval [0, 1]. The greater this coefficient, the more similar the industrial structure in the two regions. $S_{ij} = 1$ shows that the industrial structures of two regions are in full convergence; $S_{ij} = 0$ shows that the industrial structure in both regions is not convergent.

6.5 Regional Industrial Structure Policy

The target for a government making the regional industrial structure policy and implementing it is to realize some economic and social objects. Then the government places the regional industries as direct targets in making the policies and influences regional industrial distribution by protection, support, adjustment, and improvement on involved industries and participation in some operational activities of enterprises or intervention in the formation of related markets. The government aims to use the leverage of macro-economy and the means of regional industries to adjust changes of the regional industrial structure purposefully and promotes the regional industrial structure rationalization and regional economic growth. Regional industrial policy uses certain policy instruments to guide and influence the labor force division and cooperation of industrial region. It makes the resources in a reasonable allocation space. Depending on the overall evolution trend and under the premise of industrial equity and efficiency, the regional industrial policy marks out the status and functions of different regional industries in the national economic strategy. It makes industries in different regional spaces establish the reasonable regional industrial division system and the realization mechanism. It helps to exert inter-industry comparative advantages, assort regional industry configurations, improve efficiency of using resources, and promote coordinated development in regional economy. Regional industrial policy as a means and an important component of macro-adjustment mechanism is of great significance to regional economic development and rationalization of the industrial structure.*

Regional industrial structure policy one is that is made according to the changing trend of regional industrial structure to promote industrial restructuring and achieve the desired state of progress. Regional industrial structure policy plays the key support role for economic development strategy of the region. The government can support the economic development of the region, plan the goal of regional economic development and determine the future development keystone of the region's economy by using the regional industrial structure policy. For example, the government can choose the appropriate leading industries

* Jiang Qinghai. *Chinese Regional Economic Analysis.* Chongqing, China: Chongqing Publishing House, 1990.

of the region, make the regional industrial structure policy match with other regional industrial policies as well as other economic policies to help realize the goal of regional economic development effectively. Although the regional industrial structure policy cannot substitute other industrial policies and economic policies, it plays a strong supporting role to the regional economic development strategy. In the process of promoting the evolution of regional industrial structure, the pursuit of the rapid development of the regional economy has become the essence of the regional industrial structure policy.

According to the state of national economic development and regional economic development condition, the government can plan the regional industrial development and decide the key direction of the regional industrial development from a naional perspective. And the government should decide the size, speed and priority of every industrial development in the region. In accordance with the above measures, the government can guide to make regional industrial structure policies, eliminate sunset industries, support sunrise industries and then promote the advancement of regional structure effectively. If there are no regional industrial policies, and the adjustment is entirely up to the market mechanism, then the sunset industries may not be eliminated in time, and the new industries may not get the strong support and assistance. The conversion of the industrial structure may be very slow leading to a slow growth of regional economy. The use of regional industrial policies can overcome these shortcomings. Without the regional industrial policies, it would not be possible to have a reasonable regional industrial structure; moreover, there would not be appropriate regional leading industries.

Regional industrial policy should have clear objectives. This leads to the advancement of regional industrial structure, which is the total dynamic goal. During a certain period in a certain region, the regional industrial policies have their specific static goals. Connecting to different static objectives at different times, they constitute the overall dynamic objective. Therefore, the regional industrial policy is at the highest level of the industrial policy system.

The choice of leading industry is one of the most important matter in the regional industrial structure policy. Leading industries are the core and framework of the regional industrial structure; they are the "locomotive" of regional economic development. Regional industrial structure policy adjusts the regional industrial structure by chosing the leading industry sectors as well as the measures to promote their growth and development.

Regional industrial structure policy achieves the goal of restructuring the industries of the region by adjusting the supply structure as the resources supply has greater maneuverability than the market demand for a government department. If the resource supply structure is different, then its supporting industrial structure is also different. As a result, adopting a certain resources supply structure can promote the regional industrial structure to a programmed goal. At the same time, through forecasting the changing trend of the distribution structure and an analysis of the income elasticity, supply structure can also adapt to the future demand structure and improve the leading role of the regional industrial structure in the market demand.

7

Selection of Regional Leading Industry

7.1 Summarization of Selection

7.1.1 Conception and Function of Leading Industry

There is a special relationship between the industrial structure advancement and the conversion of leading industries. American economist W.W. Rostow argues that even in a mature and continually growing economy the maintenance of the impact of economic development is the result of rapid expansion of a few major growth sectors, which have a great significance for other sectors in turn.

The leading industry is the industrial sector that emerges in the course of economic development or various stages of industrialization, which has an impact on the overall situation. It takes the leading position in the national economy and promotes the economic growth through of backward, forward, and marginal correlations. It can also be called "leading specialized sector," a leading industry that has a quick development speed in the national economy and plays a leading and driving role in the overall regional economy. We can also say that the leading industry generally refers to the one that has relatively advanced production technology, high production growth rate, and strong industry conjunction. And the leading industry also occupies some weight in the overall economy and has a strong driving impact on the overall performance of other industries or the regional economy. It is the driving wheel of modern economic development. And the selection of leading industries reflects the regional industrial restructuring and optimization.

Rostow believes that the role of the leading industry can be expressed in the following three aspects.

1. Gaining of a new production function based on scientific progress

2. Forming a sustainable high growth rate
3. Having a decisive impact on the growth of other industries even all the industries with strong spillover effects

These aspects reflect the indispensable conditions of becoming the leading industries. They are an organic whole and cannot become the leading industries without any one of these aspects. According to Rostow's explanation, only few high-growth industries that have innovative abilities and proliferation effects at the same time can become leading industries.

Rostow proves that it is the role of "disproportionate growth" in the proliferation effect of the leading industries that promotes economic development. At the same time, the proliferation effects are not limited to the impact of technical and economic links among industries. It also contains a more extensive impact on economic and social development. Furthermore, with the scientific and technological progress and the development of production forces, especially the deepening of social division of labor force, it is no longer a single leading industry but several industries working together to promote the development of the whole industry. Rostow calls them "the syntheses of leading sectors." In his point of view, the syntheses of leading sectors consist of a leading sector and sectors having strong backward and marginal correlations with the leading sector. For example, iron and steel, machinery, electric power, and chemical industries together form a synthesis of leading sectors.

7.1.2 General Properties of the Leading Industry

Leading industries play a leading role in the region and usually have certain advantages or favorable properties. In terms of their specific performance in reality, most of the regional leading industries have the following properties.

7.1.2.1 Quick Growth Rate of Production From the definition of the leading industry we can see that it usually has the possibility of large-scale production. Its productivity can maintain the momentum and potential of sustainable and rapid growth so as to make the production cost decrease and to take a considerable portion in the regional GDP. In addition, it must have a higher growth rate than that of any other industries to promote the rapid growth speed of industrial and regional economy. The leading industries usually are newly emerging sectors or innovative sectors. They can introduce innovative mechanisms in relevant sectors and bring about product, technological, and institutional innovation. The high growth rate of production is associated with advanced production technology and machinery, abundant labor force resources and sufficient qualitative labor forces. Only by possessing these conditions of production, they can promote the rapid growth of industries.

7.1.2.2 High Location Quotient A high location quotient is the foundation of obtaining a comparative advantage in competition of the leading industries with industries outside of the region. It is also an important aspect of selecting the regional leading industries. It is impossible that each industry in the region has a high location quotient; therefore, the leading industries often occupy an important position in the region and have strong market forces in certain ranges. The said industry participates not only in the regional market competition, but also in the market competition outside of the region. It obtains the market position with competitive advantages and sends its products or service outside of the region. Then it gradually transfers into serving mainly the external region and then strengthens the industrial trade.

7.1.2.3 Good Market Prospect The leading industries represent the direction of industrial restructuring. Then the leading industries should have a high market demand from the point of view of equilibrium between the supply and demand. The economy is continually growing along with the increase of the income levels of the labor forces which will definitely result in changes in demand. The leading industries have to meet such demands and changes. In the meantime, they have to develop more rapidly than general industries and have greater economic increment. Therefore, it is necessary for leading industries to have relatively high income elasticity of demand and price. Namely, the industry should have a strong response of demand for changes of income.

7.1.2.4 Great Industrial Correlation The leading industries must not only pursue their own development, but also drive or promote the development of other related industries. Consequently, it is necessary for the leading industries to possess a great industrial correlation, a strong capacity of the industrial proliferation as well as a strong chain effect to become the propeller of regional economic growth. Their proliferation effect includes three forms. First is the forward effect. It is the pulling effect for the emergence of new industries, new technologies, new materials, and new energy as well as related auxiliary industries. For instance, it has a pushing impact on precision machinery, urban construction, automotive maintenance, and other related industries; the second is the backward effect; namely, the impact that the leading industries have on the development of some industries and sectors which provide raw materials for the development of leading industries such as the stimulation effect for the auto industry transferring to the iron and steel, rubber, petrochemical industry, and automobile electronics industry; the third is about marginal effect; namely, the leading industries have an impact on the development of relevant industries in the region. For example, the development of leading industries promotes the development of road construction, advertising, market industries, and so on.

7.1.3 Realization of Leading Industry

The realization of leading industry has no more than two forms according to the practice of all countries in the world.

7.1.3.1 Self-Regulation of Market The countries which take this kind of form believe that market competition and relations between supply and demand can promote the development of competitive industries. The industrial structure advancement can be achieved by supply and demand and price mechanisms. Hence, it is not necessary to establish industrial plans and develop industrial policies for the support of some leading industries. Meanwhile, they believe that the cognition of selection of leading industries by the government is no better than that of market forces. In the past, many western countries took this kind of form. They generally did not pursue the optimization of industrial structure and only gave emphasis on self-regulation of industries.

They only passively took some remedial measures and established some corresponding industrial policies when there were some problems existing in the course of economic development. However, today many countries have strengthened their intervention of the industrial development. Take the United States as an example; since 1990s it began to formulate a series of industrial technology policies to speed up the development of high-tech industries.

7.1.3.2 Active Intervention of the Government Countries select leading industries and determine the sequence of industrial development by the intervention of the government. They promote the industrial structure advancement through the continuous development of industrial policies. Japan was the earliest country that adopted this form. It did not follow the experience of European developing industries or the American developed industries, but opted for a super-conventional road of development. It selected the leading industries and determined the sequence primarily on the basis of "the benchmark of income elasticity" and "the benchmark of increased productivity."

In recent years, more and more countries are beginning to pay greater attention to the second form. Developed countries in the west get active intervention of the government to promote the development of leading industries. Being underdeveloped in the market economy, it is particularly important for developing countries to adopt policies tilted toward leading industries. In this way, they can quickly develop leading industries and continue to promote the industrial structure advancement in order to pursue the "after-effect" of "chasing" countries and the goal of sustainable and rapid development of their economies.

7.1.4 Evolution Process and Characteristics of Leading Industries

In the long history of economic development, industrial structure advancement is a historical evolution of constant replacement and conversion as well as a gradual process of industrial structure from low level to high level and from simple forms to complex forms. In this process of leading industries, there are some evolutionary characteristics.

7.1.4.1 Leading Industry Having Regional Nature In order to play the leading role the leading industry is bound to exist in a certain region or has close links with other industries. Hence, the existing of leading industries is special for the economy of a certain region. That is why there are different leading industries in different regions. The description of industrial structure can be extended to go beyond the jurisdiction of the government; however, the leading industry is specific to a certain region. Consequently, the leading industries refer specifically to the regional leading industries generally; the region includes countries or regions, as well as the smaller geographical areas in their jurisdiction. The leading role of the leading industries can be more prominent for smaller geographical areas.

7.1.4.2 Natural Process of Development of Leading Industry As Rostow said, "In essence, the economic growth after take-off can only be supported by a series of leading sectors because every leading sector is destined to slow down after a period of time." In the course of development, all industries experience the process in which they will grow from weak to strong, from young to old, and mature gradually. This is also set out by the industrial life cycle. Similarly, the leading industry is relatively weak in the initial stage of growth. With the continual development of economy, its role and proportion gradually change. In a particular period of time, the leading industries may grow quicker than other industries; then the leading industries can influence or drive the rapid development of other related industries and promote industries development through their proliferation effects. They have driving and promoting impacts on many industries. In this way, related industries continue to strengthen and grow with the economic growth; and leading industries gradually occupy the dominant positions in the overall economy. Therefore, they can play the leading role in a broader scope and establish the prominant position.

7.1.4.3 Stage Property of Leading Industries The property is reflected by two aspects. First, the leading industries always adapt to the level of economic development and have different labor productivities in different stages of economic development. The leading industries are generally different and changing along with economic development. Second, the leading industries have life cycles. With the economic development, the original leading industries will gradually enter the recession period, and the newly emerging leading industries will take over the original one to exert the leading function or the proliferation effect to promote or drive the sustainable development of the whole industries. Leading industries play their leading role only in within a certain period of time. Hence, in different stages of economic development, there are different leading industries. The leading industries of the countries or regions at different stages of development are often different. The industries that are phased out in developed countries may take rebirth in developing countries, even occupying their leading positions.

7.1.4.4 Sequence Succession of Evolution for Leading Industry In any economy the emergence of leading industries has a certain sequence of time. At different stages of economic development in a country or region, its leading industries vary. The leading industry evolves from a low-level industry to a high-level industry based on certain laws. The leading industries of different levels are substituted in an orderly manner to promote sustainable and healthy economic development. Especially in the era of industrialization, the changes of different development stages of the industrialization are shown by the orderly replacement of leading industries. In the early stage of industrialization, the developing countries are in the development phase of consumer materials based on light industry, making full use of their own comparative advantages such as resources, cheap labor, and so on in the light of the national situation. They vigorously develop the processing industries of primary products and labor-intensive industries and choose textile, food, and other light industries as the leading industries. After entering the heavy industry-centric development stage, the leading industries turn into iron and steel, energy, machine building, and chemical industries. In this stage, high processing and assembly industries, household electrical, and automobile manufacturing industry will be regarded as leading industries. With the further economic development, the process of industrialization will go through the development stage of labor-intensive to capital-intensive and then technology-intensive when such industries as aviation, aerospace, nuclear industry, computer, and other high-tech industries are selected as the leading ones in developing countries.

7.1.4.5 Multilevel of Leading Industries This property is reflected in many aspects. Yang Gongpu believes that the leading industries can be divided into two levels. The first level is to select a kind of industries as the kind of leading industries on the basis of division of three industries. Or the selection is according to the level of regional economic development and different stages of the evolution of regional industrial structure. The second level is to select some industrial sectors as the strategic sectors which will be developed vigorously on the basis of the first level.* This is an expression of focusing on different levels of industrial division. Due to the imbalanced development of regional economy, the various areas of a region may be at different stages of economic development. It will inevitably need to select different leading industries; so in a region there may be leading industries at all levels. As the substitution and evolution of leading industries is in effect, the life cycle of different industries varies. The substitution and evolution cannot happen in a split second but in a gradual process during which there is bound to be coexistence of industries at different levels. As the regional advantages of industries are indicated in different aspects, the development of regional economy also has many targets including dealing with employment pressure, enhancing industrial competitiveness, and upgrading traditional industries, etc. It is necessary to address the issue of rationalization as well as that of advancement. Accordingly, the leading industries that occupy strategic position in the regional economic development tend to be a leading industry group, which includes a combination of a variety of industries that have shown the characteristics of multilevel industries to achieve multiple objectives of regional development.

7.2 Related Theories of Selection Benchmark of Regional Leading Industries

7.2.1 Hirschman Benchmark

Hirschman, a U.S. development economist, put forward a strategy of unbalanced development in 1958 in his famous work *The Strategy*

* Yang Gongpu and Xia Dawei. *The Tutorial of Industrial Economics.* Shanghai: The Publishing Company of Shanghai Finance and Economics University, 1998, page 77.

of Economic Development. There are two ways to achieve this strategy. First is to achieve development by production shortage. Namely, industries that have high backward correlation effect should be developed quickly. The second is to achieve development by overproduction. Namely, industries that have high forward correlation effect should be developed quickly. The development policies aim to seek backward and forward correlation effects. In other words, the leading sectors can promote economic growth effectively by expanding correlation effects.

According to the basic principle of input–output, Hirschman makes further studies on the relationship between the degree of correlation among industries and the industrialization. He thinks that the developing countries which have relatively insufficient capital and small domestic market should develop industries having high correlation, especially some ultimate industries that have high backward correlation to drive the development of other industries. He also brings forward the benchmark of selecting the leading industries by the level of backward correlation. Following the experience of developing countries, Hirschman points out that there should be such an industry in the industry-correlated chain which has the highest correlation coefficient with its forward and backward industries in the relation of input and output. The development of this industry has a great potential on the forward and backward industries. He believes that the strong industrial correlation should be one of the basic guidelines to identify and select the leading industries. This means that the government chooses industries that have driving and promotional effects on many industries or have strong backward, forward, and marginal correlation as the priority for the leading industries. Hirschman's point of view has had an important impact on selecting leading industries in Japan and other developing countries.

7.2.2 Rostow Benchmark

Rostow, in his book *From Take-off into Sustainable Economic Growth* refers to Hirschman's idea of "the avail of industrial correlation" and expands the effect into the following three aspects. First, the effect of backward correlation, namely, the new sector which is in a high-speed growth will generate new input demand for raw materials and machinery and result in the rapid development of a group of industrial sectors; second, the leading sectors will cause a series of changes in the surrounding ones; third, the effect of forward correlation whereby the leading sectors promote economic development by increasing the effective supply. For example, the costs of intermediate inputs for other industrial sectors can be reduced. It shows that the Rostow's benchmark determines the leading sectors through the level of industrial correlation among industries.

The benchmark of Hirschman and Rostow determined the leading sectors through the level of industrial correlation among industries. Their focus is on the driving or promoting impacts of the leading industries. Therefore, some people put the two together as the correlation benchmark. Hirschman and Rostow only discussed the effects of industrial correlation in theory and did not measure the size of the correlation in practice.

7.2.3 Shinohara Benchmark

Japanese industrial economist Shinohara put forward the benchmark of income elasticity and the benchmark of increased productivity in the mid-1950s when he was planning the industrial structure for Japan.

The benchmark of income elasticity is the benchmark that adapts industrial structure to the growth of national income in terms of demand. Income elasticity is also known as the income elasticity of demand, which is an economic relation of the social demand for products varying with the changes in the national income under the premise of unchanged price. It is a proportion relation of the growth rate of demand of the industry and the growth rate of national income. This benchmark reflects the sensitivity of the products demand of the said industry increasing with the growth of national income. Different industries have different income elasticity of demand that shows the potential market sizes for different industries. Only the industry which has the high income elasticity of demand can continue to expand its market share. Such a kind of industry always represents the direction and trend of changes in the regional industrial structure. This benchmark emphasizes particularly on the oriented role of market demand to the growth of regional industries. Taking the income elasticity of demand as a benchmark is to give the priority to the development of industries which have larger income elasticity. And such industries have broad prospects for market development which is a prerequisite for further industrial development.

The benchmark of the productivity rate increases. In the course of economic development, there will be the inevitable inconsistency of the productivity rate increasing in various industrial sectors. It is the so-called "unbalanced growth in the productivity rate increase." Moreover, the technological progress is the main reason for improving productivity rate. If an industrial sector can make the breakthrough in technology, it will grow and develop rapidly. When income differences resulting from differences in the productivity rate increases exist among industries in the region, resources, capital, and labor forces will be transferred from low-income sectors to high-income sectors. The industrial sector making the breakthrough in technology first will obtain a rapid growth and development, and its productivity rate will maintain a high level. The relative proportion of national income created by the said industry will increase. As a pillar and driving force of the regional economic growth, this sector can promote the development of other related industries. In essence, the benchmark of the productivity rate increase reflects the benchmark of technological progress rate. In other words, it is necessary for the government to choose the industrial sector, which has a high technological progress rate as the leading one.

7.2.4 Too Dense Environment Benchmark and Rich Labor Benchmark

Both the benchmarks are of the selected leading industry. They were presented by the industrial structure council of Japan in 1971. The former benchmark requires that the industry chosen as the leading one should improve energy efficiency, strengthen the society's ability to prevent and ameliorate social effects of pollution and have a capacity to expand social capital. The benchmark indicates that when a government selects the leading industry, the following benchmarks should be followed: less environmental pollution, low energy consumption, no ecological imbalance or causing excessive harm to the environment. In fact, more and more regions choose such industries as environmental protection, ecological agriculture, tourism agriculture, and green industry as regional leading industries, which is just the application of this benchmark. This focuses mainly on the relationship between the long-term economic development and social benefit and requires the realization of socially sustainable development of regional economy.

Rich labor benchmark requires that the selection of leading industries should consider such industries which can provide workers safe and comfortable jobs and stable workplace. This standard must take full account of the richness and interest of the work content, as well as the ability to provide a safe, comfortable, and stable employment as the benchmark. This benchmark reflects that the society has finally recognized that the final objective of economic development is to improve the members' degree of satisfaction and comfort.

7.3 Principle of Selecting Regional Leading Industry

7.3.1 Maximization Principle of Income Elasticity of Demand

In the market economy, social demand is the most direct and largest impetus to promote the industrial development and changes in its structure. It is the driving force of changes and development of industrial structure. The industry which has large income elasticity of demand can create a greater market demand along with expansion of product quantity. Therefore, this type of industry can gain more development momentum from the society. And the industry can get not only its good development opportunities, but also better earnings from the development. So the speed of development for the industry will be better and the scale will be larger. The income elasticity maximization is a principle for selecting leading industries. And it is necessary for the government to choose industries whose income elasticity of demand is larger than 1 and the highest or relatively higher in the main categories as the leading industries. In this way, we can promote increasing income levels and changes in the consumption structure prominently, as well as drive the economic growth to have a broader prospect of market.

7.3.2 Maximization Principle of Rising Degree of Productivity

In the social production and the expansion of reproduction, the speed of technological progress of the industry which has larger rising degree of productivity is relatively more and the product costs are lower, generally. This need attracts a variety of resources to move toward the industry. Then there is more guarantee of supply of technology and resources than those from other industries. Consequently, it will promote the quicker development of this industry rather than other industries; a greater volume production can also be obtained; at the same time, the higher productivity rate and competitive advantage of its products at home and abroad can be achieved; and this will have greater contribution to the growth of national economy and that of the regional economy. Therefore, it is easy for this kind of industries to develop quickly. And other related industries can be promoted to develop quickly during the developing process of these kinds of industries. Then these kinds of industries can become the main pillars and key driving forces for the growth of national economy and that of the regional economy.

7.3.3 Maximization Principle of Industrial Correlation

Regional industries are complex and big which are interconnected and interdependent. Industries are different owing to various ways of production and different technological complexity. And degrees of correlation with other industries are different too. As a result, there are considerable differences in the process of promoting and driving the regional economy. There are horizontal economic ties and vertical ties of production technology among various industries. Leading industries organize and promote the development of other industries through degrees of correlation with related industries. When industries which have strong degrees of correlation develop early, there will be a series of leading and driving impacts through industrial correlation. Those industries will derive further promoting roles for other industries. Thereby, a chain reaction and accelerating effect in economic development can be obtained, which promote the overall development of the regional economy finally.

7.3.4 Principle of Labor Force Employment

As a large number of surplus labor exist in the society, so employment structure adjustment is one of the most important matter for industrial restructuring. The comprehensive employment coefficient of the labor force reflects the direct and indirect labor force input demand for this industry and other related sectors to produce one unit of final product. Thus, using this coefficient to evaluate the capacity of every industry to absorb labor force is also a principle that must be obeyed to select leading industries.

7.3.5 Principle of Dynamic Comparative Advantage

The formation and development of regional leading industries cannot be separated from the region and its objective conditions. The organic integration of the region's various conditions is the most important condition and basis of their formation and development. Therefore, the selection of regional leading industries must reflect the relative independence and characteristics. It is necessary to give full play to the geographical advantages and make full use of regional resources and conditions. The selection of regional leading industries has to be on basis of the regional reality. We must recognize that it is necessary to emphasize the play of those leading and pillar-like industries. These industries can be developed quickly based on local rich resources and advantageous conditions. Those industries whose resource advantages cannot be met in the local region should be given up. The relation of labor division and cooperation with other regions should be established well. The final goal is that leading industries in every region should develop based on their local resource advantages and features.

The selection of regional leading industries should pay more attention to not only the current regional advantages, but also the region's "latecomer advantage." They are a complex state for the economic development of the region. And this state is necessary for the regional economic development. The "latecomer advantage" refers to that advantage the region has compared with other regions with richer potential of natural, labor forces, and technical resources, and so on. The selection of regional leading industries should be based on the regional reality conditions. And the selection should also pay more attention to the development of "latecomer advantage." Only carrying out effective construction of the industry with "latecomer advantage" and efficient elements, we can continually step up other sectors of regional economy and create more vigorous social demands.

7.4 Measuring Indexes of Selecting Regional Leading Industry

At different development stages of economy and different stages of development for related industries, the parametric indexes which are required to be taken into account for leading industries will be different. According to the benchmarks and principles of selecting regional leading industries, here we classify and summarize the common measuring indexes which can be modified in accordance with local conditions while selecting leading industries.

7.4.1 Index of Industrial Correlation

Industrial correlation is divided into forward and backward correlations. Both of them refer to the degrees of input and output changes of other industrial sectors which are directly or indirectly caused by the changes in demand of an industry. There are two commonly used indexes: sensitivity coefficient and influence coefficient. Suppose b_{ij} is the coefficient of the complete consumption coefficient matrix B in the input–output analysis, n is the number of industrial sectors. Then the sensitivity coefficient μ_i of sector i is

$$\mu_i = \frac{\sum_{j=1}^{n} b_{ij}}{\frac{1}{n}\sum_{i=1}^{n}\sum_{j=1}^{n} b_{ij}}$$

μ_i indicates the extent of impact that the output increase of various sectors has on the output of sector i; it is also called forward correlation. The greater the sensitivity coefficient μ_i, the stronger the ability of forward correlation of the sector. The influence coefficient of sector j is

$$\gamma_i = \frac{\sum_{i=1}^{n} b_{ij}}{\frac{1}{n}\sum_{i=1}^{n}\sum_{j=1}^{n} b_{ij}}$$

γ_j reflects the extent of impact that the demand increase of sector j has on the output of other industrial sectors; it is also known as backward correlation. The greater the influence coefficient γ_j, the stronger the ability of the backward correlation of sector. The industrial sectors which have a larger sensitivity coefficient usually belong to basic industrial sectors, whereas the sectors which have larger influence coefficient usually belong to the final product sectors.

If γ_j exceeds 1, that means the influence of the industry on all industries is above the average; if it is equal to 1, that means the influence of the industry on all industries is average; if γ_j is less than 1, that means influence of the industry on all industries is in the lower level. In the same way, μ_i has a similar explanation.

7.4.2 *Index of Income Elasticity of Demand*

Suppose y is the gross domestic product (GDP), x_i is the product demand of sector i, then the income elasticity ε_i of sector i is given by

$$\varepsilon_i = \frac{y}{x_i}\frac{\partial x_i}{\partial y}$$

ε_i reflects the impact that changes of GDP have on the changes in the demand for products of the industrial sector i. Generally, the greater the income elasticity of demand, the stronger the ability of the the particular sector to expand the market. When ε_i is less than 1, the inflation rate of demand of sector i is lower than the rate of GDP increase; when ε_i is greater than 1, it is higher; when ε_i is equal or close to 1, it is in average level.

7.4.3 *Index of Technological Progress*

The rising productivity rate is caused by many reasons. And the most important factor is the speed of technological progress. The rate of technological progress can be derived by using the neoclassical growth model and deducting the rate of increase in the labor and capital from the production growth rate. The basic formula is as follows:

Rate of technological progress = (Growth rate of production – a)
× Labor wage growth rate – (1 – a)
× Capital growth rate

The variable a indicates the proportion of labor growth rate, assuming constant return to scale and on condition that capital and labor can substitute for each other. Utilizing b in the Cobb-Douglas production function $x_i = Ae^{bt} L_i^{\partial} K_i^{\beta}$ indicates the parameters of integrated technological progress, in which x_i is the GDP of sector i, L_i is its labor force input, k_i is its capital investment, A is a constant which reflects the level of technology. e^b is an unchanged rate reflecting the exogenous technological growth, Ae^{bt} is the level of technology, α and β indicate the contribution of labor and capital to economic growth respectively.

7.4.4 *Index of Growth*

Suppose x_i^0 is the demand for products of sector i at the initial state and r_i is the average growth rate, then the demand for products of sector i is $x_i^t = x_i^0(1+r_i)^t$. The larger the value of r_i the faster the growth of sector i and the more important its position and role in the regional economic system.

7.4.5 *Index of Employment*

Employment structure is also a major subject of industrial restructuring. The comprehensive employment coefficient is always used to estimate the absorbing capacity of labor force of every industry. Suppose v_i is the remuneration in the input–output table and q_i is the total output, c_{ij} is the coefficient of Leontif matrix, then the comprehensive employment coefficient β_i of sector i is

$$\beta_i = \sum \frac{v_i}{q_i} c_{ij}$$

β_i indicates the direct and indirect labor force input needed for this industry and other related sectors to produce one unit of final product.

7.4.6 *Index of GDP Proportion*

Suppose y_i is the GDP of sector i and y is the regional GDP. Then,

$$\alpha_i = \frac{y_i}{y}$$

α_i is the proportion of GDP of sector i in the whole regional GDP, mainly reflecting the share of contribution of various industries in the regional economy and society as well as the scale of the said industry.

8

Optimization Model of Regional Industrial Structure

8.1 Model of Estimating the Order Degree of Industrial Structure

The definition of the degree of order of industrial structure was put forward by Professor Sifeng Liu in 2004 for the first time. "It reflects the extent that a country or a region's industrial structure is close to the target structure and could be taken as a quantitative indicator to measure the level of rationalization of the national or regional industrial structure. Through the analysis of changes in order degree of the industrial structure, we can accurately grasp and judge whether or not the industrial restructuring is in the correct direction." By related theories and methods of the Grey system, Professor Sifeng Liu established the model of estimating the order degree of industrial structure and used this model to analyze and discuss the selection of objective industrial structure in China and achieved good results.

Definition 8.1. Supposing $X_i = [x_i(1), x_i(2), \ldots, x_i(n)]$ is the vector of industrial structure. Setting $x_i^0(k) = x_i(k) - x_i(1)$, then $X_i^0 = [x_i^0(1), x_i^0(2), \ldots, x_i^0(n)]$ is called the starting point image of zero.

Supposing $X_0 = [x_0(1), x_0(2), \ldots, x_0(n)]$ is the target vector of industrial restructuring, and then its starting point image of zero can be recorded as $X_0^0 = [x_0^0(1), x_0^0(2), \ldots, x_0^0(n)]$

Setting

$$|s_0| = \sum_{k=2}^{n-1} |x_0^0(k)| + \frac{1}{2}|x_0^0(n)| \tag{8.1}$$

$$|s_i| = \sum_{k=2}^{n-1} |x_i^0(k)| + \frac{1}{2}|x_i^0(n)| \tag{8.2}$$

$$| s_i - s_0 | = \sum_{k=2}^{n-1} | x_i^0(k) - x_0^0(k) | + \frac{1}{2} | x_i^0(n) - x_0^0(n) | \tag{8.3}$$

Definition 8.2. Supposing $|s_0|$, $|s_i|$, and $|s_i - s_0|$ are shown as Equations 8.1 through 8.3, respectively. Setting

$$\varepsilon_{0i} = \frac{1 + | s_0 | + | s_i |}{1 + | s_0 | + | s_i | + | s_0 - s_i |} \tag{8.4}$$

Then ε_{0i} is called the degree of order of the industrial structure for vector X_i . ε_{0i} is used to measure the extent that the vector X_i of industrial structure is close to the vector X_0 of target structure. The closer X_i is to X_0, the greater is the value of ε_{0i}. When X_i completely coincides with X_0, ε_{0i} gets the maximum value of 1.

The degree of order reflects the extent of how much the practical industrial structure is close to the objective structure. The degree of order of time series is able to indicate the direction of changes in industrial structure in a certain time of period. The direction and goal of the practical structure is the target structure, which also means that the degree of order of the practical structure can become closer to 1.

The changes in the degree of order in the industrial structure of a country or a region is shown in Figure 8.1. Figure 8.1a is an ideal process of change: the degree of order is approaching 1; besides it is increasing gradually; and Figure 8.1b shows a more general process, namely, the degree of order maybe decline during a period, but more importantly, its overall trend is rising, which reflects the correct direction of the industrial development.

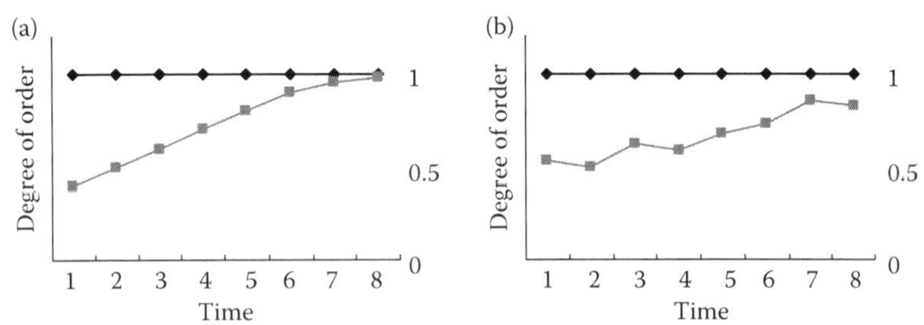

Figure 8.1 Changes in the degree of order. (a) An ideal process of change. (b) A more general process.

Before the calculation of the degree of order of the industrial structure, we should first confirm the target industrial structure. There are several methods to determine the target structure. The commonly used methods are as follows. Using the "fast track" model, which was put forward by Leontief, to solve "the track of best intensity" which can be taken as a target structure. It can also be determined in accordance with the world "standard" industrial structure of different income levels. In fact, the target structure planned by the government can be directly taken as the one for which to measure the degree of order.

8.2 "Fast Track" Model of Industrial Structure Adjustment

The economists found a law when they studied optimizing the industrial structure with input–output analysis. They found that when the studying time was long enough, the optimal solution of the output ratio of the products of various sectors in the economic system coincided with the ratio of balanced growth most of the time. In theory, the best level that the products produced are exactly the required ones. So the track of balanced growth can be considered to be the best one. In other words, if the ratio of the output in the economic system is on the track of balanced growth, it should be advancing along the track. If the economic structure is not on that track in the initial period, then it should be adjusted gradually to be on the track and then developed rapidly along the track. This theory is known as "fast track" theorem. The track of balanced growth can be calculated by the Leontief dynamic input–output model.

In the 1960s, the U.S. economist Leontief put forward the static input–output model

$$x = Ax + c \tag{8.5}$$

where

$x = [x_1 \;\; x_2 \;\; \ldots \;\; x_n]^T$ is the vector of total output.

$c = [c_1 \;\; c_2 \;\; \ldots \;\; c_n]^T$ is the vector of final demand.

$A = \begin{bmatrix} a_{11} & \cdots & a_{1n} \\ \vdots & & \vdots \\ a_{n1} & \cdots & a_{nn} \end{bmatrix}$ is the matrix of direct consumption coefficient.

As for the matrix A, $a_{ij} = x_{ij}/x_j$, x_{ij} is the number of products of sector i which are intermediate input for sector j. In other words, a_{ij} indicates that a_{ij} units of products of sector i are needed to produce one unit of the products of sector j.

Although the Leontief static input–output model takes into account the usage and needs of fixed assets, it does not separate the structure of investment and consumption. Therefore, there are some flaws in this theory. In order to solve these problems, Leontief broke down the use of fixed asset into investment and consumption structure and established the dynamic input–output model on the basis of the static input–output model as shown:

$$x(t) = Ax(t) + D[x(t+1) - x(t)] + c(t) \tag{8.6}$$

where

$$D = \begin{bmatrix} d_{11} & \cdots & d_{1n} \\ \vdots & & \vdots \\ d_{n1} & \cdots & d_{nn} \end{bmatrix}$$

D is the coefficient matrix of the use of fixed assets; d_{ij} is the value of fixed assets of sector i used to produce the products of sector j per 10,000 Yuan, that is to say, d_{ij} = (the total value of fixed assets produced by sector i in sector j)/(total output of sector j). $c(t) = [c_1(t)\ \ c_2(t)\ \ldots\ c_n(t)]^T$ is the vector of final consumption demand of the year t.

The basic idea of Leontief's dynamic input–output model is that under conditions of equilibrium, the total output $x(t)$ is applied to the demand of intermediate inputs $Ax(t)$, demand of investment $D[x(t+1) - x(t)]$ as well as the demand of final consumption $c(t)$, and there is an equilibrium equation as follows:

The total output = Intermediate inputs + Consumption + Investment

As for the vector of final consumption demand, the formula is given by

Consumption structure = Average consumption of per unit of working hours
× The consumption proportion of every sector
× The total input labor force hours

The average consumption of per unit of working hours is given by $\dfrac{C_1+C_2+\cdots+C_n}{L_1+L_2+\cdots+L_n}$, where C_i is the consumption of sector i and L_i is the inputs' working hours for sector i. The consumption proportion of every sector is $\left(\dfrac{C_1}{C},\dfrac{C_2}{C},\ldots,\dfrac{C_n}{C}\right)^T$ where C is the total consumption of various sectors.

The total input labor forces hours is $L_1+L_2+\cdots+L_n=l_1x_1(t)+l_2x_2(t)+\cdots+l_nx_n(t)$. Then $l_i=\dfrac{L_i}{x_i(t)}$ indicates the number of labor force hours required for the output of per unit products of sector i. Therefore, the consumption structure $c(t)$ is given by

$$c(t)=\frac{C_1+C_2+\cdots+C_n}{L_1+L_2+\cdots+L_n}\begin{pmatrix}\dfrac{C_1}{C}\\\dfrac{C_2}{C}\\\vdots\\\dfrac{C_n}{C}\end{pmatrix}\left(\frac{L_1}{x_1(t)},\frac{L_2}{x_2(t)},\ldots,\frac{L_n}{x_n(t)}\right)\begin{pmatrix}x_1(t)\\x_2(t)\\\vdots\\x_n(t)\end{pmatrix}=Tx(t)$$

where T is the consumption coefficient matrix.

Thereby, we can get another expression of the dynamic input–output model as shown:

$$x(t)=Ax(t)+D[x(t+1)-x(t)]+Tx(t) \tag{8.7}$$

Equation 8.7 is the Leontief "fast track" model.

In practice, the economic growth rates of various sectors are inconsistent. For example, the growth rate of agriculture sector may be relatively slower than that of other industries due to the limitation of natural resources and regional conditions such as land, climate, and so on. However, in the dynamic input–output model shown in Equation 8.7, the growth of various sectors belong to extensive economic growth because it did not consider the resource constraints and other complex factors as well as technology progress. In this case, theoretically, the economic growth in various sectors can be considered to be consistent. If the growth rates are inconsistent, namely, that some sectors grow with high growth rate and some with low rate, then the gap of output levels will be very large when time t is large enough. It is assumed that the economic growth of all sectors is consistent, then,

$$x(t+1) = (1+\alpha)x(t) \tag{8.8}$$

Inserting Equation 8.8 into the dynamic input–output model (Equation 8.7), the following equation is derived:

$$\begin{aligned} x(t) &= Ax(t) + D[(1+\alpha)x(t) - x(t)] + Tx(t) \\ &= Ax(t) + \alpha Dx(t) + Tx(t) \end{aligned}$$

$$\frac{1}{\alpha}x(t) = (I - A - T)^{-1}Dx(t) \tag{8.9}$$

If we let $(I - A - T)^{-1}\ D = H$, then the above Equation 8.9 can be rewritten as

$$\frac{1}{\alpha}x(t) = Hx(t) \tag{8.10}$$

From Equation 8.10 we know that the equilibrium growth rate α of Leontief dynamic input–output model is the reciprocal of the latent root $1/\alpha$ of matrix H, the output vector $x(t)$ is the corresponding unit eigenvalue of this latent root. The unit eigenvalue can be recorded as $x(t) = [x_1(t), x_2(t), \dots x_n(t)]^T$. If the actual industrial structure is not on the track of balanced growth, and then it should be adjusted to be on the "fast track" as soon as possible. However, the regional economic structure is gradually shaped after a long-term evolution, so there is a large inertia. It is impossible for the industrial structure adjustment to be carried out for a very short time. It should be a gradual process. Therefore, it is generally assumed to adjust the industrial structure to be on the "fast track" in a certain period. Here we assume to adjust it to be on the "fast track" after m years and record the total production values of all industries in that year as $x(m) = [x_1(m)q, x_2(m)q, \dots x_n(m)q]^T$, where q is the undetermined coefficient. The process of adjustment shall make not only the industrial structure on the track of balanced growth after m years but also the total output as large as possible, namely, to maximum q.

Considering that the total demand must be less than the total output every year, the constraint is as follows:

Demand for intermediate input $AX(t)$ + Demand for investment $D[X(t+1) - X(t)]$ + Demand for final consumption $TX(t) \leq$ Output $X(t)$

Based on the above analysis, the linear programming model is available as

$$\max q = \begin{cases} AX(0) + D[X(1) - X(0)] + TX(0) \leq X(0) \\ AX(1) + D[X(2) - X(1)] + TX(1) \leq X(1) \\ \cdots \\ AX(m-1) + D[X(m-1) - X(m-1)] \\ \quad + TX(m-1) \leq X(m-1) \\ \bar{X} \cdot q \leq X(m) \\ X(1), X(2), \ldots, X(m) \geq 0, \quad q \geq 0 \end{cases} \tag{8.11}$$

where, A, D, and T are mentioned as above. $X(0)$ is the industrial structure at the initial period, $\bar{X}$ is the optimal structure, $X(1)$, $X(2)$, … , $X(m)$ are the industrial structures from the initial year to the mth year, respectively and are all n-dimensional row vectors. Using the software LINGO to solve the linear programming model mentioned above, the industrial structure corresponding to the production value of all industries in the year m can be calculated. Then it can be changed into the one that is calculated as gross domestic product in accordance with certain conditions.

8.3 Grey Dynamic Linear Programming Model of Industrial Structure Adjustment

In the course of industrial structure adjustment, if the input–output table of this region is available, we can use the "fast track" model which is introduced in last section to carry out industrial structure adjustment and optimization. However, in general, this table is not available for the city located in the region. In this case we cannot use the "fast track" model, so we construct the grey dynamic linear programming model of industrial structure adjustment. For convenience, it is assumed that when studying the industrial structure adjustment in this region, we only adjust the three major industries.

The industrial structure optimization refers to the process of coordinated development of various industries and the continual improvement of the overall development level of the industries. The essence of industrial structure optimization is to achieve the optimal allocation and efficient usage of resources among industries, and then to promote a coordinated, stable, and efficient economic development. The industrial structure optimization includes not only the structure of added values but also employment structure among the three major industries. As the formation of the industrial structure optimization results from the flow of production factors such as labor force, capital, energy, natural resources, and technological progress among various industries, the establishment of the mathematical model of industrial structure and employment structure optimization would also be done from the same flow. The natural resources are indicated by the demand for water in the region. However, it is difficult to draw the contribution and quantification of scientific and technological progress in various industries. Hence, when building the mathematical model, we do not take into account the role of scientific and technological progress in various industries. We take these resources as constraints and establish the grey dynamic linear programming model to optimize the industrial structure.

Supposing x_1, x_2, and x_3 are the GDP of the primary, secondary, and tertiary industry, respectively (a hundred million); the grey dynamic linear programming model is established as follows:

$$\max Z = x_1 + x_2 + x_3$$

$$\begin{cases} a_{11}x_1 + a_{12}x_2 + a_{13}x_3 \leq b_1 \\ a_{21}x_1 + a_{22}x_2 + a_{23}x_3 \leq b_2 \\ a_{31}x_1 + a_{32}x_2 + a_{33}x_3 \leq b_3 \\ a_{41}x_1 + a_{42}x_2 + a_{43}x_3 \leq b_4 \\ x_1, x_2, x_3 \geq 0 \end{cases}$$

where b_1 represents the labor force input in this region (10,000 people); b_2 represents the capital input in the region (10,000 yuan); b_3 represents the energy input (10,000 tons of standard coal); and b_4 represents the water demand in the region (10,000 tons). a_{11}, a_{12}, and a_{13} refer to the quantity of labor force demand for output value of 10,000 yuan of the primary, secondary, and tertiary industry, respectively, in this region (people/10,000 yuan); a_{21}, a_{22}, and a_{23} refer to the quantity of capital demand for output value of 10,000 Yuan of the primary, secondary, and tertiary industry, respectively, in this region (10,000 yuan/10,000 yuan); a_{31}, a_{32}, and a_{33} refer to the quantity of energy demand for output value of 10,000 yuan of the primary, secondary, and tertiary industry, respectively, in this region (10,000 tons of standard coal/10,000 yuan); a_{41}, a_{42}, and a_{43} show the quantity of water demand for output value of 10,000 Yuan of the primary, secondary, and tertiary industry, respectively, in this region (10,000 tons/10,000 yuan).

The grey dynamic linear programming model is established as follows:

$$\max Z = x_1 + x_2 + x_3$$

$$\begin{cases} a_{11}x_1 + a_{12}x_2 + a_{13}x_3 \le b_1 \\ a_{21}x_1 + a_{22}x_2 + a_{23}x_3 \le b_2 \\ a_{31}x_1 + a_{32}x_2 + a_{33}x_3 \le b_3 \\ a_{41}x_1 + a_{42}x_2 + a_{43}x_3 \le b_4 \\ x_1, x_2, x_3 \ge 0 \end{cases}$$

Let us predict labor force input b_1, capital input b_2, energy input b_3 and water demand b_4 in this region respectively. Utilizing the data from 1999 to 2005, we can construct the GM(1,1) model to predict b_1, b_2, b_3, and b_4 from the year 2006 to 2010.

As for a_{11}, a_{12}, a_{13}, a_{21}, a_{22}, a_{23}, a_{31}, a_{32}, a_{33}, a_{41}, a_{42}, a_{43}, we can use data from the year 1999 to 2005 to construct GM(1,1) model and predict resources consumption coefficients from the year 2006 to 2010.

Then the grey dynamic linear programming models from 2006 to 2010 can be constructed and solved to get the optimal industrial structure each year.

$$\max Z_{2006} = x_1 + x_2 + x_3$$

$$\begin{cases} a_{11}^{2006} x_1 + a_{11}^{2006} x_2 + a_{11}^{2006} x_3 \leq b_1^{2006} \\ a_{21}^{2006} x_1 + a_{22}^{2006} x_2 + a_{23}^{2006} x_3 \leq b_2^{2006} \\ a_{31}^{2006} x_1 + a_{32}^{2006} x_2 + a_{33}^{2006} x_3 \leq b_3^{2006} \\ a_{41}^{2006} x_1 + a_{42}^{2006} x_2 + a_{43}^{2006} x_3 \leq b_4^{2006} \\ x_1, x_2, x_3 \geq 0 \end{cases}$$

....

$$\max Z_{2010} = x_1 + x_2 + x_3$$

$$\begin{cases} a_{11}^{2010} x_1 + a_{11}^{2010} x_2 + a_{11}^{2010} x_3 \leq b_1^{2010} \\ a_{21}^{2010} x_1 + a_{22}^{2010} x_2 + a_{23}^{2010} x_3 \leq b_2^{2010} \\ a_{31}^{2010} x_1 + a_{32}^{2010} x_2 + a_{33}^{2010} x_3 \leq b_3^{2010} \\ a_{41}^{2010} x_1 + a_{42}^{2010} x_2 + a_{43}^{2010} x_3 \leq b_4^{2010} \\ x_1, x_2, x_3 \geq 0 \end{cases}$$

Through the optimization of industrial structure, the employment structure can be calculated by multiplying x_i (the output value each year is calculated by the model) by b_i (the consumption coefficient of labor of various industries each year forecasted by the model).

8.4 Grey Clustering Decision-Making Model of the Selection of Regional Leading Industries

According to related theories, leading industries refer to the industrial sectors which play leading and driving roles in the industrial structure and economic development at a certain stage of economic development of a country or region and have broad market prospects and the capacity for technological progress. In order to determine the regional leading industries, the status and role of a sector in regional economic development must be reflected by the combined effects of a number of indexes. Consequently, we can make use of the grey clustering model to establish the model of selecting regional leading industries.

Grey clustering analysis is one of the main components of grey system theory and is also a way of clustering analysis which means classifying the targets studied in accordance with a certain grey class definition. A clustering can be seen as the gathering of observation targets belonging to the same type. Practically, each observation object often has many indexes of feature. Therefore, it is difficult to make accurate classification. Based on the whitened weight function of the number of grey, the grey fixed weight clustering is used to classify some observation indexes or targets into a number of defined categories. It is mainly used to check whether the observation targets belong to different grey types that have been set in advance so as to treat them differently.

Definition 8.3. Supposing there are n targets of clustering, m indexes of clustering and s different grey types. The index value of clustering target i on the clustering index j is x_{ij} $(i = 1, 2, \ldots, n; j = 1, 2, \ldots, m)$.

The whitened weight function of the clustering index j on k grey type is given by $f_j^k(*)$ $(j = 1, 2, \ldots, m; k = 1, 2, \ldots, s)$.

1. Supposing λ_j^k is the threshold value of index j and sub-category k, then the expression:

$$w_j^k = \frac{\lambda_j^k}{\sum_{j=1}^{m} \lambda_j^k}$$

can be called the weight of index j and sub-category k, and

$$\sigma_i^k = \sum_{j=1}^{m} f_j^k(x_{ij}) w_j^k$$

is called the clustering coefficient of grey changed weights of target i on index k.

2. If the weight of clustering of the index j on the grey type k has nothing to do with k, namely, that w_j $(j = 1, 2, \ldots, m)$ is the weight of clustering of the index j and $\sum_{j-1}^{m} w_j = 1$, then $\sigma_i^k = \sum_{j=1}^{m} f_j^k(x_{ij}) w_j$ can be called the grey fixed weight clustering coefficient of target i belonging to the grey type k.

Definition 8.4. If $\sigma_i = (\sigma_i^1, \sigma_i^2, \ldots, \sigma_i^s)$ $(i = 1, 2, \ldots, n$ is known as the clustering coefficient vector of clustering target i, then,

$$\Sigma = (\sigma_i^k) = \begin{bmatrix} \sigma_1^1 & \sigma_1^2 & \cdots & \sigma_1^s \\ \sigma_2^1 & \sigma_2^2 & \cdots & \sigma_2^s \\ \cdots & \cdots & \cdots & \cdots \\ \sigma_n^1 & \sigma_n^2 & \cdots & \sigma_n^s \end{bmatrix}$$

Σ is the matrix of grey clustering coefficient.

Definition 8.5. Supposing $\max_{1 \le k \le s}\{\sigma_i^k\} = \sigma_i^{k^*}$, we can say that target i belongs to the grey type k_*.

According to the theory of grey system clustering analysis, the concrete steps of establishing the model of regional leading industries selection are as follows.

Supposing there are n targets, m evaluation indexes and s different grey types. The evaluation value of target i on index j is x_{ij} $(i = 1, 2, \ldots, n; j = 1, 2, \ldots, m)$.

Step 1. Based on the related theory mentioned above, we can determine the index set of selecting the regional leading industries. Supposing there are n industries, m evaluation indexes of regional leading industries. In accordance with the characteristics and requirements of selecting leading industries, n industries can be divided into s different grey types. The observation value of target i on the index j is x_{ij} $(i = 1, 2, \ldots, n; j = 1, 2, \ldots, m)$. In terms of types of selecting the regional leading industries, the regional industries are generally classified into leading industries, supporting industries, and common industries three types.

Step 2. Determining the various types of whitened weight function $f_j^k(*)$ $(j = 1, 2, \ldots, m;\ k = 1, 2, \ldots, s)$ of each index as well as the weight w_j $(j = 1, 2, \ldots, m)$ of various indexes in the selection of leading industries through Delphi method.

Step 3. Calculating the grey clustering coefficient σ_i^k of target i on the grey type k, and $\sigma_i^k = \sum_{j=1}^m f_j^k(x_{ij})w_j$.

Step 4. Determining that target i belongs to grey class k^* by $\max_{1 \le k \le s}\{\sigma_i^k\} = \sigma_i^{k^*}$.

If the number of leading industries calculated is relatively large, we need to sort them so as to determine the leading industries.

In the process of decision-making, we may experience the fact that a number of decision-making targets belong to the same grey type of decision-making which can only accommodate limited targets. In such a case, we need to sort these decision-making targets in order to make choices. As the vectors of grey clustering coefficient are not unit vectors, they cannot be compared with each other. Therefore, we do some changes for grey clustering coefficient as follows.

Definition 8.6. Let $\delta_i^k = \sigma_i^k / (\sum_{k=1}^s \sigma_i^k)$. It can be called the integrated clustering coefficient belonging to the target i on grey type k. Then,

$$\Pi = \delta_i^k = \begin{bmatrix} \delta_1^1 & \delta_1^2 & \cdots & \delta_1^s \\ \delta_2^1 & \delta_2^2 & \cdots & \delta_2^s \\ \cdots & \cdots & \cdots & \cdots \\ \delta_n^1 & \delta_n^2 & \cdots & \delta_n^s \end{bmatrix}$$

Π is called the matrix of integrated grey clustering coefficients.

Definition 8.7. In terms of $\max_{1 \le k \le s} \left\{ \delta_i^k \right\} = \delta_i^{k^*}$, we can say that the cluster target i belongs to the grey class k^*.

Proposition 8.1. Determining the decision-making target i by the clustering coefficient σ_i^k or integrated clustering coefficient δ_i^k, the two results are the same; namely, for the decision-making targets, they are in the same grey type.

Although the integrated clustering coefficient has been unitized in the same grey type, due to the difference of forward and backward data among its subvectors, it has still a great impact on the order of various targets. For example, if $\delta_1 = (0.4, 0.38, 0.22)$ and $\delta_2 = (0.41, 0.2, 0.39)$, both the two targets belong to the first category in terms of the direct comparison of their integrated clustering coefficient; since $0.41 > 0.4$, target 2 should be ranked prior to target 1. However, the sum of the coefficients of target 1 in the first and second type of grey is much more than that of target 2, and the clustering coefficient of target 2 is close to that of target 1 in the first grey type. So one would normally think that the clustering result of target 1 should be ranked before that of target 2. This results in the differences between the results and one's intuition; the reason is that when we compare the integrated clustering coefficients, we do not consider the coefficients as a whole.

If a target belongs to the grey type k, we have enough reason to think that the sum of clustering coefficients of the mentioned target in the grey type k. Since the target belongs to the grey type k, we can consider that its weight of clustering coefficient in that type is the biggest and the farther it is from the grey type k, the smaller is its corresponding weight.

Definition 8.8. Supposing the decision-making target has s different grey types. Let

$$\eta_1 = (s, s-1, s-2, \ldots, 1)$$

$$\eta_2 = (s-1, s, s-1, s-2, \ldots, 2)$$

$$\eta_3 = (s-2, s-1, s, s-1, \ldots, 3)$$

$$\ldots$$

$$\eta_k = (s-k+1, s-k+2, \ldots, s-1, s, s-1, \ldots, k)$$

$$\eta_{s-1} = (2, 3, \ldots, s-1, s, s-1)$$

$$\eta_s = (1, 2, 3, \ldots, s-1, s)$$

and $\eta_1, \eta_2, \ldots, \eta_s$ are respectively called the adjustment coefficient of the first, second, and so on of the first s grey type.

Note: η_k ($k = 1, 2, \ldots, s$) is the s-dimensional vector where the first k sub-vector is s. Taking s as a center, the sub-vectors decline on both sides with pace 1.

Definition 8.9. Supposing there are n decision-making targets, s different grey types, if target i is in the grey type k, then we call $\omega_i = \eta_k \delta_i^T$ its comprehensive decision-making measure.

Definition 8.10. Supposing $\max\limits_{1\le k\le s}\{\delta_{i_1}^k\} = \delta_{i_1}^{k^*}$, $\max\limits_{1\le k\le s}\{\delta_{i_2}^k\} = \delta_{i_2}^{k^*}$, target i_1 and i_2 both belong to grey type k^*. If $\omega_{i_1} > \omega_{i_2}$, we say that i_1 is better than i_2 in k^*.

Definition 8.11. Supposing $\max\limits_{1\le k\le s}\{\delta_{i_1}^k\} = \delta_{i_1}^{k^*}$, $\max\limits_{1\le k\le s}\{\delta_{i_2}^k\} = \delta_{i_2}^{k^*}$, … and $\max\limits_{1\le k\le s}\{\delta_{i_l}^k\} = \delta_{i_l}^{k^*}$, which means target i_1, i_2, …, i_l all belong to k^* and $\omega_{i_1} > \omega_{i_2} > \cdots > \omega_{i_l}$. If the number of the target which could be accommodated by k^* is l_1, then i_1, i_2, …, i_{l1} are targets selected by k^*, the rest are backup targets for k^*.

In summary, we can obtain the arithmetic of the grey integrated clustering model as follows:

1. Classify the s grey types in accordance with the requirements of integrated clustering, and the ranges of various targets should be correspondingly divided into s grey types to determine the whitened weight function $f_j^k(*)$ ($j = 1, 2, \ldots, m$; $k = 1, 2, \ldots, s$).
2. Identify the clustering weights of each target according to the conclusions of qualitative analysis.
3. Calculate the clustering coefficient σ_i^k of target i on the grey type k, by using $\sigma_i^k = \sum_{j=1}^{m} f_j^k(x_{ij})w_j$, where $f_j^k(x_{ij})$ is the whitened weight function belonging to the grey type k of target i under the index j, w_j is the weight of index j in the decision-making of grey evaluation.
4. Calculate the integrated clustering coefficient δ_i^k of target i on the grey type k, and $\delta_i^k = \sigma_i^k / \sum_{k=1}^{s} \sigma_i^k$.
5. If $\max\limits_{1\le k\le s}\left\{\delta_i^k\right\} = \delta_i^{k^*}$, we can determine that target i belongs to k^*.
6. Using all targets belonging to the grey type k to calculate their integrated clustering measure $\omega_i = \eta_k \delta_i^T$.
7. Sort all the targets of grey type k according to the comprehensive clustering measure ω_i and select the leading industries based on the given limited number.

9

Research on Industrial Structure Optimization and Upgrading in M City, W Province during the Period of "11th Five-Year Plan"

Located on the south bank of the lower reaches of the Yangtze River, M City is a node city on the axis side of the Yangtze River Delta, which is the eastern part of W Province. In recent years, trying to achieve the goal of joining the Yangtze River Delta with an accelerating speed and getting a first breakthrough in the development of Wan Jiang project, the city has maintained a good momentum of rapid and healthy economic development. Over the years, many of the key per capita economic indexes of the city such as gross domestic product (GDP), revenue, income, and savings for urban residents ranked first in the W Province. In 2004, its GDP amounted to 26,509 million RMB. Calculated in accordance with comparable price, M City had a GDP growth of 19.5% over the previous year, which is the largest increase in W Province.

Over 20 years of development since the reform and opening to the outside world M City has obtained many initial achievements in optimizing the industrial structures of the three major industries, which show all-round, coordinated, and healthy development momentum. The increment of the primary industry in GDP has been declining, which has dropped to 6.8% in 2004; the proportion of the secondary industry has had a large fluctuation, reaching 69.3% in 2004; the share of the tertiary industry decreased, after increase at first, only 23.9% in 2004. The employment proportion of M City in the primary and secondary industries has shown a downward trend, while that of the tertiary industry has been rising gradually. The employment proportion of the secondary industry apparently declined from 1990 to 2003, with a decrease of 5.76%; at the same period, the share of the tertiary industry increased 9.94%. The tertiary industry has become the main channel of M City for absorbing the labor force.

9.1 Status Quo and Problems in the Industrial Structure of M City

9.1.1 Analysis of the Internal Structure of the Three Major Industries in M City

9.1.1.1 Analysis of the Internal Structure of the Primary Industry The primary industry is divided into four sectors, which are agriculture, forestry, animal husbandry, and fisheries, respectively. We can study the changes in the internal structure of the primary industry of M City in terms of the output value of the four sectors.

From Table 9.1 we find that the internal structure of the primary industry on the whole reflects a reduction in the proportion of agriculture, an increase in fisheries, a fluctuating rise in forestry, and a downward trend in animal husbandry in recent years after rise in the early years. Among the four sectors, the proportion of agriculture and fisheries has a relatively larger range of changes; the proportion of forestry is smaller and the proportion of animal husbandry changes in a relatively smaller range. At present, the fisheries have become a sector with the largest proportion in the primary industry of M City. This shows that the internal structure adjustment in the primary industry of M City has made encouraging achievements. Although the proportion of traditional agriculture has been declining, the fisheries has made a rapid development through giving full play to its advantages of fresh water resources.

Table 9.1 Production Value Structure of Agriculture, Forestry, Animal Husbandry, and Fisheries in M City (%)

PRIMARY INDUSTRY	1978	1980	1985	1990	1995	2000	2003
Agriculture	81.21	73.95	76.59	69.49	54.11	44.95	38.02
Forestry	1.13	1.83	1.10	0.73	1.71	1.43	1.73
Animal husbandry	14.33	18.89	15.94	19.69	18.23	16.20	16.27
Fisheries	3.33	5.33	6.47	10.09	25.96	37.42	43.08

Source: Statistical Yearbook of M City, 2004.

Table 9.2 Internal Production Value Structure of the Secondary Industry in M City (%)[a]

SECONDARY INDUSTRY	1995	1996	1998	2000	2002	2003
Mining	6.16	5.62	5.35	4.61	3.41	3.59
Food and tobacco processing	2.24	2.36	1.39	2.13	4.14	3.85
Textile clothing and leather	4.80	5.05	5.43	5.73	3.75	3.55
Wood processing, paper-making industry and the cultural and educational supplies	1.94	1.86	1.90	3.18	3.15	4.60
Petroleum refining, chemical, and pharmaceutical	5.00	4.15	3.59	2.97	2.71	3.21
Nonmetallic mineral products	1.00	1.22	1.89	1.40	0.89	1.04
Metal smelting and fabricated metal products	66.39	66.38	67.26	66.24	62.44	58.79
Machinery	4.82	4.49	4.00	3.70	3.11	3.75
Transportation equipment manufacturing	1.33	1.07	1.62	3.31	10.26	11.71
Electrical and electronic equipment manufacturing	4.04	2.85	2.49	2.25	2.27	2.10
Public utilities	2.28	4.95	5.08	4.48	3.87	3.81

Source: Statistical Yearbook of M City, 1996, 1997, 1999, 2001, 2003, 2004.

[a] Proportion of the production value of each sectbr in the total industrial production value.

9.1.1.2 Analysis of the Internal Structure of the Secondary Industry The changing trend reflected in Table 9.2 shows that the proportion of industrial production value in some industries declined, including the mining industry, textile clothing and leather industry, metal smelting and fabricated metal products industries, machinery industry, petroleum refining, chemical, pharmaceutical industry, electrical and electronic equipment manufacturing industry; at the same time, some other industries had a rising trend; these were food and tobacco processing industry, wood processing and paper-making industry and the cultural and educational supplies, transportation equipment manufacturing, and public utilities.

Although the proportion of industrial output value in metal smelting and fabricated metal products industry declined from 66.39% to 58.79%, lowering by 7.6%, it still accounted for more than half of the total industrial production value of M City, while the transportation equipment manufacturing sector climbed from 1.33% to 11.71% bringing an increase of 10.38%. The rise of industries such as transportation equipment manufacturing industry as well as the decline in the proportion of the industrial output value in metals and fabricated metal products industry describe that M City has gradually nurtured a number of new industries; and the condition of single industrial structure in M City has improved gradually. However, the metal smelting and fabricated metal products industry still accounts for a proportion of 58.79% in the total industrial production value. So the single industrial structure in secondary industry is still serious, in fact.

9.1.1.3 Analysis on the Internal Structure of the Tertiary Industry The changing trend reflected in Table 9.3 shows that the proportion of increment in the tertiary industries such as geological survey and water resources management basically maintained stable from 1997 to 2002, while some other industries such as transportation, warehousing, posts and telecommunications had a downward trend every year and decreased significantly. Comparatively, the industries such as finance and insurance, real estate, and social services increased significantly from 1997 to 1999; there was a 7.03% increase; although it decreased every year from 1999, the proportion still increased by 3.3% in 2003 compared to that in 1997.

9.1.2 Problems in the Industrial Structure of M City

M City has made many achievements in economic development and industrial structure adjustment, but some problems in the industrial structure are still found through the analysis on the industrial structure of the three major industries and the internal structure of each main industry. The major problems are discussed below.

Table 9.3 Internal Structure of the Tertiary Industry in M City (%)

TERTIARY INDUSTRY	1997	1998	1999	2000	2002	2003
Geological survey and water resources management	1.57	1.74	1.65	1.49	1.49	2.16
Transportation, warehousing, and posts and telecommunications	24.83	21.47	20.60	19.85	19.53	16.31
Wholesale and retail trade catering	24.23	18.72	18.28	22.09	21.22	22.13
Finance and insurance, real estate and social services	35.46	41.24	42.49	40.32	39.96	38.76
Culture, education, health, and scientific research	9.23	11.19	11.07	10.54	11.64	11.64
Others[a]	4.68	5.64	5.91	5.71	6.16	9.00

Source: Statistical Yearbook of M City, 1998, 2000, 2002, 2004.

[a] Others include physical education and social welfare, state organs, organs of political parties and social groups, as well as other tertiary industries.

9.1.2.1 High Proportion of Employment in the Primary Industry The proportion of employment in the primary industry still accounted for 35.29% in 2003. The high proportion of employment impeded the effective improvement in the labor productivity of the primary industry. From 1990 to 2003, the GDP proportion of the primary industry fell faster than the decline in the proportion of employment in M City, which explains the decline in the comparable labor productivity of the primary industry. This situation will further affect the rural economic development and rural market development, resulting in gradual expansion of the difference between urban and rural area, which is not conducive to the harmonious development of the whole society.

9.1.2.2 Single Internal Industrial Structure In the internal structure of heavy industry, raw material industry, which is represented by metal smelting and fabricated metal products industry, took up a very large scale in the industrial output value, as high as 70.69% in 2003; while the proportion of processing industry was too low. Raw materials industry is at the low-end of the industrial chain with a low added value, and it also contributes toward greater environmental damage. The effect made by economic cycles on raw material industry is more obvious, which is also the reason why the internal structure of the secondary industry of M City fluctuated. With raw material industry as the leading industry of M City, it is inevitable that the economic development depends on resources excessively. In the increasingly heated environment of global scramble for resources, this kind of dependence will increase the instability of the economic system.

9.1.2.3 Serious Lack of Development in the Tertiary Industry The development of the tertiary industry is too slow in the performance of the low proportion of the tertiary industry in the industrial structure. The high proportion of the secondary industry of M City is not caused by excessive development, but by the slow development in the primary industry and tertiary industry, especially the latter. We also find that the development of the tertiary industry of M City is relatively slower than that planned in the 10th Five-Year Plan, and the level of its internal structure is low. Both the structure of employment and of production value of the tertiary industry are apparently on a low level.

9.1.2.4 Low Industrial Technology Level and Added Value The economic development of M City is mainly supported by the traditional industries. Few emerging technology industries and short industrial chain further lead to the low added value of the industries. The serious shortage of investment in science and technology also limits the industrial upgrading of M City at a certain extent. The three items of sci-tech funds accounted for only 1.18% in the fiscal expenditure, which is lower than the national average level of 1.96%.

9.2 Main Influence Factors of the Industrial Structure Upgrading in M City

There are many complex factors resulting in the evolution of the industrial structure in M City that are subjective and objective as well as internal and external. From a macro point of view, these factors mainly include science and technology, investment structure, labor force, industrial policy, natural resources, changes in demand, and other factors. Further elaboration of the impacts of these factors on the industrial structure upgrading in M City are discussed below.

9.2.1 Impact of Science and Technology

Science and technology are the primary productive forces, and technological innovation is the fundamental driving force for economic development. Science and technology development is an important factor to promote optimization and upgrading of industrial structure. From 2001 to 2003, M City significantly increased the investment in science and technology, spending 13.02 million RMB, 15.26 million RMB, and 20.4 million RMB, respectively, in the three items of sci-tech funds for a fiscal year, with respective growth of 27.77%, 17.2%, and 33.7% compared with the number of the previous year. However, the three items of sci-tech funds still took up a low rate of fiscal expenditure, which were 1.28%, 1.14%, and 1.18%, respectively. The in-house expenditures of the large and medium-sized industrial enterprises of the city in scientific and technological activities were 442 million RMB and 798 million RMB, and 1302 million RMB, with respective growth of 46.7%, 80.5% and 63.2% compared with the amount of the previous year. The instability in the professional technical personnel employed was presented in the state-owned economic units and urban collective economic units, the steady increase existed only in other types of economic units, but the total number of professional technical personnel employed showed a downward trend, as shown in Table 9.4.

Technological progress will play an increasingly important role in the process of the industrial structure optimization and upgrading in M City. Take Maanshhan Steel Company (Ma Steel) as an example; the company engages in ongoing technological transformation and product restructuring; on the one hand, Ma Steel tries hard to increase the capacity and efficiency by drawing out the potential of the existing equipment to improve its efficient usage; on the other hand, it speeds up the development of high-tech and high value-added products in order to guarantee the rapid development in the period of 10th Five-Year Plan. The formation of the industrial clusters in the business of magnetic materials, represented by Tian Yuan Company, marks the start and development of the high-tech industries in the city.

The optimization and upgrading of the industrial structure in M City needs to reinforce the technological transformation in traditional industries, especially in large and medium-sized enterprises, to increase investment in science and technology so that the total investment not only increases but also has an increasing proportion in the fiscal expenditure, and at the same time, the structure of the investment is further optimized. Furthermore, it needs to nurture high-tech industries and seek a new growth point for the economic development of M City.

9.2.2 Impact of Investment Structure

Risk is one of the important factors that leads to changes in the industrial structure, and the alterations in either the total investment or

Table 9.4 Number of Professional Technical Personnel Employed in Different Types of Economic Units in M City

ECONOMIC UNITS	2001	2002	2003
State owned	29,347	29,407	28,192
Urban collective	3,206	2,624	2,703
Other types	7,800	7,847	8,812
Total	40,353	39,878	39,707

Table 9.5 Accumulated Investment in the Infrastructure of M City from 2000 to 2003 (million RMB)

THREE INDUSTRIES	2000	2001	2002	2003
Primary industry	77.93	55.96	39.43	12.10
Secondary industry	3285.85	1018.10	897.69	1177.38
Tertiary industry	1121.08	1340.97	2205.38	2566.85

Source: Statistical Yearbook of M City, 2001, 2002, 2003, 2004.

investment structure would give rise to changes in the industrial structure (Table 9.5).

In 2003 as a whole, the accumulated investment in fixed assets of M City marked 10,000 million RMB for the first time, reaching 10,679 million RMB and increasing 73.1% compared to that in 2002. For the full year of 2004, the total investment of fixed assets amounted to 14,546 million RMB, having a growth of 36.2% more than that of the previous year. The investment structure has also been improved. At the same time, M City has completed the investment of 90 million RMB in the primary industry, 2.6 times that in 2003, an investment of 9853 million RMB in the secondary industry with a growth of 32.2% compared to that in 2003, and an investment of 4425 million RMB in the tertiary industry, 48% more compared to that in 2003.

From the data of the accumulated investment in infrastructure of M City, it is easy to find that the infrastructure investment in the primary industry declined in 2003, the accumulated amount of investment is less than one-sixth of that in 1999; the change in the infrastructure investment in the secondary industry was unstable; the infrastructure investment in the tertiary industry was on the rise, the accumulated investment in 2003 was 2.3 times than that in 1999.

M City has also made some achievements in attracting foreign investment, having played a positive role in promoting the optimization and upgrading of industrial structure. The absolute amount of actual utilized foreign direct investment in M City was basically on the rise from 1999 to 2003. The foreign investment of foreign-funded enterprises that signed new agreements comprised 0.74% in the primary industry, 88.43% in the secondary industry and 10.83% in the tertiary industry. The investment in the primary industry was totally for the agricultural investment. The investment in the secondary industry was mainly for manufacturing sector investment, accounted for 80.42% of the total foreign investment, followed by electricity, production gas and water supply industry, and took up 8.02% of the total. In the tertiary industry, investment was concentrated on the real estate industry, water conservancy, environment, and public facilities management industry with respective shares of 7.00% and 3.82% in the total foreign investment.

During the period of 11th Five-Year Plan, it was necessary for M City to attract domestic and foreign capital actively to guide the optimization of investment structure, guide more capital into the primary industry and tertiary industry, especially to enhance investment of infrastructure in both the primary industry and tertiary industry and rely on adjustment of the investment direction to promote the optimization and upgrading of industrial structure.

9.2.3 *Impact of Labor Force*

The number, quality, and the direction of flow of labor force have a direct impact on the changes of the industrial structure. The industries with labor force that has certain quality flow will get the conditions for developing. From the number of employees in the three major industries of M City from 1999 to 2003, we find that there was a trend of flow of the labor force from the primary and secondary industries to the tertiary industry. The number of employees in the primary industry showed a decreasing trend on the whole from 1999 to 2003; the number of employees in the secondary industry declined each year; while the number in the tertiary industry began to increase annually from 1999, with an average annual growth of 3.03% over the previous year (Table 9.6).

The structure of employees in three major industries has been in the continuous optimization from 1999 to 2003, but the secondary industry took up the largest share, followed by the primary industry,

Table 9.6 Number of Employees in Three Major Industries of M City from 1999 to 2003 (in Thousands)

THREE INDUSTRIES	1999	2000	2001	2002	2003
Primary industry	253.0	246.0	247.0	235.0	223.0
Secondary industry	257.0	254.0	251.0	247.0	239.0
Tertiary industry	151.0	152.0	158.0	160.0	170.0

Table 9.7 Structure of Employees in Three Major Industries of M City from 1999 to 2003 (%)

THREE INDUSTRIES	1999	2000	2001	2002	2003
Primary industry	38.28	37.73	37.65	36.61	35.28
Secondary industry	38.88	38.96	38.26	38.47	37.82
Tertiary industry	22.84	23.31	24.09	24.92	26.90

and then the tertiary industry. Generally speaking, a large number of workers in the primary industry of M City need to be transferred, as the employment absorption capacity of the secondary industry has dropped, the problem could be solved only through the rapid development of the tertiary industry. Therefore, during the period of 11th Five-Year Plan, it is necessary for M City to develop the tertiary industry as a focus of industrial structure adjustment to solve the employment problem of M City (Table 9.7).

9.2.4 Impact of Industrial Policies

In the market economy, the industrial policy orientation of the government plays an important role in the optimization and upgrading of industrial structure. M City put forward the industrial policy in its 10th Five-Year plan of the national economy and social development to speed up the implementation of projects affecting the adjustment of agricultural structure, one of which was to promote the economic development of forestry industry through the implementation of "the construction of the production base for the industrial raw material forest," "the project of bamboo industry in Xiang Shan district," and "the project of fruit forest," and so on. The total output value of the forestry was on the rise in a straight line, with an average annual growth of 18.12%. The 10th Five-Year Plan of the development of the tertiary industry also proposed strengthening the building of tourist facilities to develop the tourism industry vigorously, focusing on the introduction of three brands, the natural landscape represented by Cai Shiji, the human landscape represented by the cluster of tombs of Libai and Zhuran, and the urban landscape featured by gardens with mountains and waters and steel production, and striving to become an important tourism node in East China. In 2004, the revenue of foreign exchange of M City in tourism reached US$ 8,372,500, more than 2.2 times revenue compared to that in 2001. M City set up a number of district-level industrial parks in 2002, leading to the rapid development of the processing industry. This shows that the impact of the industrial policies in the optimization and upgrading of industrial structure is enormous.

Overall, the implementation of industrial policy of M City during the period of the 10th Five-Year Plan vigorously promoted the rapid development of related industries. During the period of 11th Five-Year Plan, M City will continue to implement these effective policies, and at the same time, it is necessary to develop new industrial policies to support the development of leading industries, to support the development of high-tech industries, to support the development in industrial technology level, to develop a new industrial system, and finally, to promote the optimization and upgrading of industrial structure.

9.2.5 Impact of Natural Resources

As a resource-based industrial city, the formation and development of M City is closely related to its natural resource conditions. M City, next to the Yangtze River Delta area, has convenient water, land, and air traffic, and is rich in natural resources. All of these provide a well-suited condition for its industrial structure adjustment.

M City is rich in mineral resources. There are 31 places of proven iron ore production, with the total iron ore reserves of 1635 million tons, taking up 57.32% of the total iron ore reserves of W Province. There are some great potential reserves of mines, such as the village of Gao, the village of Tao, Heshang Bridge, which are located in the outskirts of M City, and White Elephant Mountain in Dangtu County. Sulfur iron ore mines are concentrated in the outskirts of the city such as the Xiang Mountain, Ma Mountain area, with a total reserve of about 262 million tons, accounting for about 55.39% of the reserves of W Province. M City also has abundant high-grade phosphor reserves, about one-third of the reserves of M Province. Potassium ore is an important raw material of ceramic, glass and paper industry, having a reserve of one million tons, and is mainly distributed in the west of Geyang Mountain, suburb of M City. Besides, the potassium ore here has a good possibility for mining because of its thin stripping layers. Alum ore which is the raw material of potash fertilizer and sulfuric acid is mainly distributed in the large Yellow Mountain in the area of

Xiang Mountain, whose reserve is about 2.1 million tons, with an alum grade of 38.7%. Gypsum used as one of the ingredients in the cement industry is located in Xiang Mountain in the southeast of the urban areas. In addition, there are some other mineral reserves for exploitation; they are nonferrous metals such as gold, copper ore, and nonmetallic minerals such as Gao ling clay, mica, and so on.

The abundant mineral resources of M City have made great contributions to the development of the city. The economic development of the city is to a large extent dependent on its mineral resources. However, we find from the current situation and development trend of M City that it has been very difficult for local resources to meet the needs of the development of the city, which need introduction of external resources in a large scale. For example, 70% of the iron ore for Ma steel needs to be imported from abroad, and along with the development of Ma Steel, the ratio will continue to increase. With the optimization of the industrial structure of M City and the advance of its technological progress as a whole, the local mineral resources will have decreasing impact on the optimization and upgrading of the industrial structure.

Natural conditions of M City such as climate, geographical location still cannot be ignored in the industrial structure adjustment. M City is located in the Yangtze River Delta area, next to Nanjing, having a certain advantage, which will help the city to carry on the domestic and international industrial transfers and share the regional division of labor. The facilitated transportation offers necessary conditions for economic, scientific, technological, cultural, and social exchanges and cooperation of M City and its surrounding areas. Less per capita cultivated land and rich fresh water resources determine that the internal adjustment of the city in primary industry will inevitably lead to an increase in the proportion of fisheries and a decline in agriculture.

9.2.6 Impact of Changes in Demand

Market demand determines both the existence value of all the economic activity and the necessity of the existence of a particular industry. As a result, changes in demand are the direct cause leading to changes in the industrial structure. With the economic development, the structure of demand will have corresponding stage changes and show administrative levels and orderly evolution, becoming the basis for the evolution of the industrial structure.

Consumer demand of M City is undergoing profound changes. Income of urban and rural residents has increased in a large range, urban per capita disposable income of M City in 2004 marked 10,000 RMB for the first time, reaching 10,189 RMB, with an increase of 15.9% over the previous year; per capita net income of the farmers increased 863 RMB over the previous year, reaching 3985 RMB, with a growth of 27.6% over the previous year. It is inevitable that with the great improvement in the income of urban and rural residents of the city, further enhancement in the purchasing power of residents, and continuous changes in demand structure will bring along optimization of the industrial structure, especially for attracting the tertiary industry.

9.3 Selection of Leading Industries in M City

9.3.1 Index System of the Selection of Leading Industries in M City

Regions are at different developing stages; as a result, the actual situation of the region should be comprehensively taken into consideration in the selection of leading industries. M City did not compile the input–output table, so we could not get the inter-industry relationship. Therefore, when choosing the leading industries of M City, we consider the actual situation of the city and choose the following indicators as the index system of the selection of its leading industries.

1. *Income elasticity of demand:* Suppose Y is the GDP, x_i is the demand for the product of sector i; then the income elasticity of demand of sector i is given by $\varepsilon_i = (y/x_i)(\partial x_i/\partial y)$.
2. *Growth rate:* Suppose x_i^0 is the demand for the product of sector i in the initial state, r_i is the average growth rate, and then the demand for the product of sector i in the period t is: $x_i^t = x_i^0(1 + r_i)^t$. The bigger the r_i value, the faster the growth of sector i and its developing speed; thus, the status and role of r_i will become increasingly important in the regional economic system.

3. *Employment absorption rate:* In industrial structure adjustment, the employment structure is also a major subject of restructuring. In the selection of leading industries, the employment absorption should be given to full consideration. Employment absorption rate = annual average employment of an industry/total assets of the industry. This indicator shows the contrast relations between the capital and labor force in the process of production, namely, employment opportunities provided by a certain amount of investment. We can calculate the amount of investment needed for providing a job by analyzing the indicator, which reflects the job function of the leading industries.
4. *Labor productivity:* Labor productivity is a concentrated expression of the level of the industrial production. Technological progress is an important reason for improving productivity, the industry sector that has the first breakthrough in technology will grow and develop rapidly, and its productivity will be able to maintain a high rate of increase, and then, the relative proportion of national income created by this industrial sector will increase. In the selection of leading industries, the inter-sector labor productivity should also be given full consideration. The formula for calculation of labor productivity is as follows: Labor productivity = increment of an industry in a year/annual average employment of the industry. The improvement in productivity is specifically embodied in the rate of technological progress; therefore, the sector with rapid improvement in productivity has a higher rate of technological progress, which is also the candidate sector for the leading industries.
5. *Growth rate of labor productivity:* Growth rate of labor productivity is growth speed of labor productivity. Labor productivity and growth rate of labor productivity can reflect the technology content and the technological progress of an industry to a certain extent.
6. *Proportion of an industry's increment in GDP:* This is given by the formula: Proportion of an industry's increment in GDP (%) = (the increment of an industry in a year/GDP of the year) × 100. Proportion of an industry's increment in GDP can reflect whether or not an industry of a region has a certain

foundation, as well as the contribution of the industry to economic growth in the region.

9.3.2 *Selection of Leading Industries in M City*

According to the *Statistical Yearbook* of M City over the years, we can use the grey fixed clustering model of the selection of regional leading industries to choose the regional economic leading industries and auxiliary industries of M City.

1. The data indicators are calculated on the basis of *Statistical Yearbook* of M City over the years.
2. In order to facilitate the calculation, departments are merged to 19 departments. The results are shown in Table 9.8.

All noted evaluation indicators are dimensionless indexes; therefore, in order to facilitate the evaluation, the regional departments are divided into three different types, which are leading industries, auxiliary industries, and general industries, in accordance with the importance in the regional economic development. Through the Delphi survey, we ascertained the various types of whitened weight function and the weight in the general assessment of each indicator. The results are shown in Table 9.9.

The selection results of leading industries, which are eight leading industries, five auxiliary industries, and six general industries, are shown in Table 9.10.

Leading industries are those of food and tobacco processing, wood processing, paper-making industry and the cultural and educational supplies, metal smelting and fabricated metal products, transportation equipment manufacturing, wholesale, and retail trade catering.

Auxiliary industries are of petroleum refining and chemical, pharmaceutical, nonmetallic mineral products, machinery, electrical and electronic equipment manufacturing, finance and insurance, real estate, social services, culture, education, health and scientific research and others.

General industries are those of agriculture, mining, industry, textile, clothing and leather, public utilities, construction, geological survey and water resources management, water management, transportation, warehousing, posts and telecommunications.

Table 9.8 Evaluation Index Value of Each Industrial Sector of M City

INDUSTRY	INCOME ELASTICITY OF DEMAND	GROWTH RATE (%)	EMPLOYMENT ABSORPTION RATE	LABOR PRODUCTIVITY (10,000 RMB/ PERSON)	GROWTH RATE OF LABOR PRODUCTIVITY (%)	PROPORTION OF AN INDUSTRY'S INCREMENT IN GDP (%)
Agriculture	0.37	1.56	1.64	0.61	5.56	7.38
Mining	1.17	4.01	0.22	4.63	7.73	3.97
Food and tobacco processing	2.62	18.96	0.05	18.99	19.13	0.75
Textile clothing and leather industry	0.76	3.99	0.14	7.40	5.10	0.82
Wood processing, paper-making industry and the cultural and educational supplies	2.22	33.43	0.02	55.13	39.08	0.97
Petroleum refining and chemical, pharmaceutical	1.47	13.35	0.05	19.30	41.87	0.81
Nonmetallic mineral products	1.73	13.99	0.10	9.66	24.97	0.20
Metal smelting and fabricated metal products	1.59	15.20	0.03	34.40	26.44	39.53
Machinery	1.53	9.12	0.09	11.47	29.48	0.81

Transportation equipment manufacturing	2.83	60.84	0.01	74.99	42.72	0.87
Electrical and electronic equipment manufacturing	1.48	9.03	0.06	16.22	31.09	0.19
Public utilities	1.03	5.25	0.03	29.14	9.37	5.96
Construction	1.08	6.59	0.42	2.24	6.39	8.37
Geological survey and water resources management	1.19	8.87	0.51	4.13	22.48	0.45
Transportation, warehousing, posts and telecommunications	1.12	2.83	0.33	3.83	7.53	6.04
Wholesale and retail trade catering	1.40	8.63	0.69	1.44	11.73	6.50
Finance and insurance, real estate, and social services	1.54	11.94	0.11	8.72	4.54	11.38
Culture, education, health, and scientific research	1.69	15.75	0.29	3.15	11.22	3.60
Others	1.85	23.01	0.27	2.33	27.21	1.40

Note: The labor productivity in the table is the data from 2003. The growth rate of the labor productivity is the average data from 2000 to 2003.

Table 9.9 Grey Whitened Weight Function and Its Weight

	GREY CATEGORY OF INDUSTRY			
INDEX	GENERAL	AUXILIARY	LEADING	WEIGHT
Income elasticity of demand	$f_1^1(-,-,0.8,1.5)$	$f_1^2(0.8,1.5,-,2)$	$f_1^3(1,2,-,-)$	0.17
Growth rate	$f_2^1(-,-,5,10)$	$f_2^2(5,10,-,20)$	$f_2^3(10,20,-,-)$	0.13
Employment absorption rate	$f_3^1(-,-,0.05,0.1)$	$f_3^2(0.05,0.1,-,0.35)$	$f_3^2(0.1,0.4,-,-)$	0.15
Labor productivity	$f_4^1(-,-,3,10)$	$f_4^2(3,10,-,20)$	$f_4^3(8,24,-,-)$	0.18
Growth rate of labor productivity	$f_5^1(-,-,5,10)$	$f_5^2(5,15,-,35)$	$f_5^3(15,35,-,-)$	0.12
Proportion of an industry's increment in GDP	$f_6^1(-,-,0.4,1)$	$f_6^2(0.5,1,-,6)$	$f_6^3(1,10,-,-)$	0.25

Clustering coefficients in various types of grey category of each industry are greater than zero, which indicates that each industry has both advantages and disadvantages. Leading industries, auxiliary industries, and general industrial may transform into each other. For example, industries like machinery industry, culture, education, health, and scientific research industry have similar clustering coefficients in leading industries and auxiliary industries. This indicates that these kinds of industries have a large possibility of transforming into leading industries from auxiliary industries and can also be developed vigorously.

The selection of leading industries is actually an orientation of the regional division of labor and industrial layout of M City in the Yangtze River Delta region and even the whole country. If the industrial structure and layout are briefly copied from that of other regions, it is probable that the leading industries will not play their due role, and some related industries will develop slowly as a result of competition and other factors. Therefore, it is necessary to compare the selected leading industries with those of other cities in the Yangtze River Delta region (see Table 9.11).

We find that some industries here are more or less repetitions of those in other regions, such as transportation equipment manufacturing, machinery processing, and other industries. But they cannot be understood as simple repeat of the industries. On the one hand, due

Table 9.10 Evaluation Results of Industries in M City

INDUSTRY	INTEGRATED CLUSTERING COEFFICIENT (GREY CATEGORY OF INDUSTRY)			MAX	CLUSTERING RESULTS INDUSTRY
	GENERAL	AUXILIARY	LEADING		
Agriculture	0.58656	0.00672	0.327208	0.58656	General
Mining	0.402709	0.343988	0.171424	0.402709	General
Food and tobacco processing	0.252376	0.254069	0.434898	0.434898	Leading
Textile clothing and leather	0.558195	0.401858	0.02	0.558195	General
Wood processing, paper-making industry and the cultural and educational supplies	0.160526	0.237369	0.6	0.6	Leading
Petroleum refining and chemical, pharmaceutical	0.236806	0.41634	0.370575	0.41634	Auxiliary
Nonmetallic mineral products	0.258743	0.551367	0.254465	0.551367	Auxiliary
Metal smelting and fabricated metal products	0.15	0.25316	0.66654	0.66654	Leading
Machinery	0.132915	0.527538	0.416018	0.527538	Auxiliary
Transportation equipment manufacturing	0.20488	0.184144	0.6	0.6	Leading
Electrical and electronic equipment manufacturing	0.350077	0.391423	0.270615	0.391423	Auxiliary
Public utilities	0.402763	0.116983	0.322775	0.402763	General
Construction	0.4573	0.12602	0.3684	0.4573	General
Geological survey and water resources management	0.485286	0.299511	0.22718	0.485286	General
Transportation, warehousing, posts and telecommunications	0.440223	0.141417	0.275271	0.440223	General
Wholesale and retail trade catering	0.239906	0.320854	0.370648	0.370648	Leading
Finance and insurance, real estate and social services	0.152914	0.552266	0.38012	0.552266	Auxiliary
Culture, education, health and scientific research	0.176143	0.39513	0.359282	0.39513	Auxiliary
Others	0.162	0.390948	0.313756	0.390948	Auxiliary

Table 9.11 Comparison of Leading Industries between M City and Cities in the Yangtze River Delta Region

AREA	LEADING INDUSTRIES
Shanghai	Electronic information product manufacturing, automotive manufacturing, petrochemical and fine chemical manufacturing, steel products manufacturing, complete sets of equipment manufacturing industry, biomedical manufacturing
Hangzhou	Machinery, electronics, food, textile and garment, medicine and chemical
Jiaxing	Textile, clothing, leather, warp, wood furniture, magnetic materials, glass fiber and new components
Huzhou	Textiles, nonmetallic mineral products, electrical machinery and equipment manufacturing, wood processing and wood, bamboo and brown grass products, electric and thermal power production and supply, the fabricated metal products, pharmaceutical manufacturing
Zhoushan	Ports, fisheries, seafood processing industry, tourism
Ningbo	Electronic information, textile and garment, household appliances, machinery and mold, construction materials
Shaoxing	Textiles, chemicals, electronics industry
Taizhou	Auto and parts, household electrical appliances, medicine and chemical, the plastic mold, garment machinery
Nanjing	Electronics, automotive, petrochemical, iron and steel, electric power
Suzhou	Electronic information, precision machinery and automotive spare parts
Wuxi	Electronic information and software, automobile and mechanical and electrical integration, top-grade textiles and clothing, new materials, biomedicine
Changzhou	Agricultural machinery, power transmission equipment, motor vehicles (motorcycles, engineering vehicles, locomotives) and its accessories manufacturing, the new type of textile service
Yangzhou	Electronic appliances, biomedicine, textiles and clothing, ships, chemicals, mechanical equipment, processing of agricultural products, automobiles, tourism
Nantong	Fine chemicals, ship repair, textiles and garments, marine products and processing, electronics, machinery
Zhenjiang	Paper-making, chemical, metal tools, auto accessories, electrical engineering, and so on
Taizhou	Electrical machinery and equipment manufacturing, general equipment manufacturing, medicine manufacturing, chemical materials and chemical products, transportation equipment manufacturing, textilc, fabricated metal products
M City	Food and tobacco processing industry, wood processing, paper-making industry and the cultural and educational supplies, metal smelting and fabricated metal products industry, transportation equipment manufacturing, wholesale and retail trade catering industry

Source: Economic and Social Development Statistical Bulletin of these cities.

to the shortage of characteristic resources and the coarse division of the industries, there are few optioned industries with development potential and few choices, so the repetitions are inevitable; however, on the development of content, transportation equipment manufacturing industry of M City relies currently on Xingma Group which manufactures heavy-duty truck products; therefore, the apparent conflict between M City and other regions do not exist. Machinery processing industry covers more content. The development of cutting and moldings tools is comparatively mature in machinery processing industry of M City, which also does not have conflict with other areas. As a whole, essentially, there are not many conflicts in leading industries between M City and the Yangtze River Delta region.

9.4 Industrial Structure Optimization of M City

Industrial structure optimization includes increment structure and employment structure of the three major industries. The industrial structure optimization is mainly formed by the flow of production factors in different industries, such as labor force, capital, energy, natural resources, and technological progress. Therefore, taking these resources as constraints, we establish a grey dynamic linear programming model to calculate the industrial structure optimization of M City.

9.4.1 Optimization of the Industrial Structure and Employment Structure of Three Major Industries in M City during the Period of 11th Five-Year Plan

Although the industrial structure and employment structure of M City have been optimized to a certain extent, the problem of structure optimization still has not been fundamentally resolved. In order to seek the optimal structure of the three major industries of M City, we established the grey dynamic linear programming model. As it is difficult to get quantitative contribution of the scientific and technological progress in various industries, the effect of this progress in each industry will not be considered in the process of establishing the mathematical model.

Supposing x_1, x_2, x_3 are the GDP of the primary, secondary, and tertiary industries (a hundred million RMB), b_1 is the labor force input (10,000 people); b_2 is the capital investment (10,000 RMB); b_3

is the power input (10,000 degrees); b_4 is the water demand (10,000 tons). a_{11}, a_{12}, a_{13} are the labor force required for production value of 10,000 RMB of the primary, secondary and tertiary industries, respectively (people/10,000 RMB); a_{21}, a_{22}, a_{23} are the funds required for production value of 10,000 RMB of the primary, secondary, and tertiary industries respectively (10,000 RMB /10,000 RMB); a_{31}, a_{32}, a_{33} are the electrical energy required for production value of a hundred million RMB of the primary, secondary, and tertiary industries respectively (10,000 degrees/10,000 RMB); a_{41}, a_{42}, a_{43} are amount of water required for production value of a hundred million RMB of the primary, secondary, and tertiary industries, respectively (10,000 tons/10,000 RMB); then we establish the following grey dynamic linear programming model:

$$\max Z = x_1 + x_2 + x_3$$

$$\begin{cases} a_{11}x_1 + a_{12}x_2 + a_{13}x_3 \le b_1 \\ a_{21}x_1 + a_{22}x_2 + a_{23}x_3 \le b_2 \\ a_{31}x_1 + a_{32}x_2 + a_{33}x_3 \le b_3 \\ a_{41}x_1 + a_{42}x_2 + a_{43}x_3 \le b_4 \\ x_1, x_2, x_3 \ge 0 \end{cases}$$

We predict b_1, b_2, b_3, b_4, respectively. Based on the data from 1999 to 2003, we construct GM(1,1) model separately to predict the constraint volume of each year from 2006 to 2010.

Then predict a_{11}, a_{12}, a_{13}, a_{21}, a_{22}, a_{23}, a_{31}, a_{32}, a_{33}, a_{41}, a_{42}, a_{43}, respectively. Based on the data from 1999 to 2003, we construct GM(1,1) model separately to predict resource consuming coefficient of each year from 2006 to 2010.

The GDP and a variety of data of M City from 1999 to 2003 are shown in Tables 9.12 through 9.16.

According to the above data, we use the grey system theory to establish GM(1,1) model of each indicator above, obtaining the predicted value of a variety of consuming coefficients and resource constraints. The specific results are given in Tables 9.17 through 9.19.

According to the above predicted value of consuming coefficients and the total amount, we construct grey dynamic linear programming model of structure optimization of three major industries of M City

Table 9.12 GDP of M City from 1999 to 2003 (10,000 RMB)

YEAR	PRIMARY INDUSTRY	SECONDARY INDUSTRY	TERTIARY INDUSTRY
1999	124,227	774,344	337,931
2000	126,383.5	796,366.5	386,275.5
2001	128,540	818,389	434,620
2002	131,753	929,536	484,638
2003	135,743	1,249,531	539,963

Table 9.13 Number of Employees and the Labor Force Required for Production Value of 10,000 RMB of M City from 1999 to 2003

	NUMBER OF EMPLOYEES IN THE INDUSTRY (10,000 PEOPLE)				LABOR FORCE REQUIRED IN THE INDUSTRY (PEOPLE/10,000 RMB)		
YEAR	PRIMARY	SECONDARY	TERTIARY	TOTAL	PRIMARY	SECONDARY	TERTIARY
1999	25.3	25.7	15.1	66.1	2.0365943	0.3318938	0.4468368
2000	24.6	25.4	15.2	65.2	1.9464566	0.3189486	0.3935015
2001	24.7	25.1	15.8	65.6	1.9215808	0.3067001	0.363536
2002	23.5	24.7	16	64.2	1.7836406	0.265724	0.3301433
2003	22.3	23.9	17	63.2	1.6428103	0.1912718	0.3148364

Table 9.14 Total Investment and the Funds Required for Production Value of 10,000 RMB of M City from 1999 to 2003

	TOTAL INVESTMENT (10,000 RMB) IN THE INDUSTRY				FUNDS REQUIRED IN THE INDUSTRY (10,000 RMB/10,000 RMB)		
YEAR	PRIMARY	SECONDARY	TERTIARY	TOTAL	PRIMARY	SECONDARY	TERTIARY
1999	5538	20,248	57,206	82,992	0.0445797	0.0261486	0.1692831
2000	1213	37,110	48,117	86,440	0.0095978	0.0465991	0.1245665
2001	5596	34,493	61,869	101,958	0.0435351	0.0421474	0.1423519
2002	2493	46,871	138,171	187,535	0.0189218	0.0504241	0.2851015
2003	1210	101,001	157,100	259,311	0.0089139	0.0808311	0.2909459

Table 9.15 Consumption of Electricity and the Electrical Energy Required for Production Value of 10,000 RMB of M City from 1999 to 2003

	CONSUMPTION OF ELECTRICITY IN THE INDUSTRY (10,000 DEGREES)				ELECTRICAL ENERGY REQUIRED IN THE INDUSTRY (10,000 DEGREES/10,000 RMB)		
YEAR	PRIMARY	SECONDARY	TERTIARY	TOTAL	PRIMARY	SECONDARY	TERTIARY
1999	7148	265,601	33,464	306,213	0.0575398	0.3430013	0.0990261
2000	7986	290,735.5	31,303.5	330,025	0.0631886	0.3650775	0.0810393
2001	8824	315,870	29,143	353,837	0.0686479	0.3859656	0.067054
2002	8688	350,020	33,259	391,967	0.0659416	0.3765535	0.0686265
2003	8271	402,147	38,523	448,941	0.0609313	0.3218384	0.0713438

Table 9.16 Consumption and Amount of Water Required for Production Value of 10,000 RMB of M City from 1999 to 2003

	CONSUMPTION OF WATER IN THE INDUSTRY (10,000 TONS)			AMOUNT OF WATER REQUIRED IN THE INDUSTRY (10,000 TONS/10,000 RMB)	
YEAR	SECONDARY	TERTIARY	TOTAL	SECONDARY	TERTIARY
1999	39,976	4957	44,933	0.0516256	0.0146687
2000	40,705	4554	45,259	0.0511134	0.0117895
2001	41,434	4151	45,585	0.0506287	0.0095509
2002	41,193	3988	45,181	0.0443157	0.0082288
2003	42,111	4100	46,211	0.0337014	0.0075931

Source: Statistical Yearbook of M City.

Note: Due to the lack of individual statistical yearbook, we use instead the average of two years (the years before and after the said year). The amount of water consumption in the *Statistical Yearbook* of M City includes water consumption for living and total water supply only, the water supply in rural areas has not been covered; so the water consumption of the primary industry has been omitted when we dealt with this case.

Table 9.17 Predicted Value of Labor Force, Investment Funds, Electric Power and Water Consumption

YEAR	EMPLOYEES (10,000 PEOPLE)	INVESTMENT FUNDS (100 MILLION RMB)	ELECTRIC POWER (10,000 DEGREES)	WATER CONSUMPTION (10,000 TONS)
2005	62.016449	558,104.35	546,317.13	46,427.146
2006	61.312618	828,213.49	606,913.72	46,678.484
2007	60.616774	1,229,049	674,231.57	46,931.182
2008	59.928828	1,823,879.2	749,016.21	47,185.249
2009	59.248689	2,706,593.1	832,095.84	47,440.691
2010	58.576269	4,016,519.3	924,390.53	47,697.516

Table 9.18 Predicted Value of the Labor and Investment Funds Required for Production Value of 10,000 RMB

	LABOR REQUIRED IN THE INDUSTRY (PEOPLE/10,000 RMB)			INVESTMENT FUND REQUIRED IN THE INDUSTRY (10,000 RMB/10,000 RMB)		
YEAR	PRIMARY	SECONDARY	TERTIARY	PRIMARY	SECONDARY	TERTIARY
2005	1.4918242	0.1584437	0.2662018	0.0149065	0.1161228	0.5574005
2006	1.4095206	0.1365121	0.246374	0.0136633	0.1458306	0.7474318
2007	1.3317577	0.1176163	0.228023	0.0125237	0.1831384	1.0022493
2008	1.258285	0.101336	0.2110389	0.0114792	0.2299908	1.3439402
2009	1.1888657	0.0873092	0.1953199	0.0105218	0.2888295	1.8021219
2010	1.1232763	0.0752239	0.1807716	0.0096442	0.3627208	2.4165086

Table 9.19 Prediction Value of Electrical Energy and Water Required for Production Value of 10,000 RMB

	ELECTRICAL ENERGY REQUIRED IN THE INDUSTRY (10,000 DEGREES/10,000 RMB)			WATER REQUIRED IN THE INDUSTRY (10,000 TONS/10,000 RMB)	
YEAR	PRIMARY	SECONDARY	TERTIARY	SECONDARY	TERTIARY
2005	0.0615354	0.3182533	0.0625256	0.0288152	0.005323
2006	0.0606684	0.3067456	0.0600722	0.0254503	0.0045618
2007	0.0598136	0.2956539	0.057715	0.0224784	0.0039094
2008	0.0589709	0.2849634	0.0554504	0.0198535	0.0033504
2009	0.05814	0.2746594	0.0532746	0.0175351	0.0028712
2010	0.0573208	0.2647279	0.0511842	0.0154875	0.0024606

from 2005 to 2010, respectively. For example, the grey linear programming model in 2005 is as follows:

$$\max Z_{2005} = x_1 + x_2 + x_3 e$$

$$\begin{cases} 1.4918242x_1 + 0.1584437x_2 + 0.2662018x_3 \leq 620164.49 \\ 0.0149065x_1 + 0.1161228x_2 + 0.5574005x_3 \leq 558104.35 \\ 0.0615354x_1 + 0.3182533x_2 + 0.0625256x_3 \leq 546317.13 \\ 0.0288152x_2 + 0.005323x_3 \leq 46427.146 \\ x_1, x_2, x_3 \geq 0 \end{cases}$$

By calculating the grey linear programming model, we can get the optimal solution (unit: 10,000 RMB, the same as below)

$$x_1 = 138924, \quad x_2 = 1488305, \quad x_3 = 665288.4$$

In the same way, we can construct grey linear programming model of structure optimization of the three major industries of M City from 2006 to 2010, respectively, and calculate the optimal solutions.

According to the results of the model, we can obtain the structure optimization program of the three major industries of M City during the period of 11th Five-Year Plan on annual GDP basis, and then get the employment structure optimization program of M City during same period, in accordance with the predicted value of the labor force required for production value of 10,000 RMB and structure optimization program of the three major industries for every year deduced from grey dynamic linear programming model as shown in the Table 9.20.

Table 9.20 Structure Optimization Program of Three Major Industries of M City during the Period of 11th Five-Year Plan (%)

	PROPORTION					
	PRIMARY INDUSTRY		SECONDARY INDUSTRY		TERTIARY INDUSTRY	
YEAR	GDP	LABOR	GDP	LABOR	GDP	LABOR
2005	6.06	33.42	64.92	38.02	29.02	28.56
2006	5.69	33.28	64.99	36.78	29.31	29.94
2007	5.20	32.43	65.06	35.82	29.74	31.75
2008	4.86	32.03	64.87	34.46	30.28	33.50
2009	4.56	31.74	64.63	33.04	30.80	35.22
2010	4.34	31.64	63.97	31.21	31.69	37.15

We find from Table 9.20 that through the optimization during the period of 11th Five-Year Plan, the proportion of GDP in the industrial structure of the three major industries will be adjusted from 6.06, 64.92, and 29.02 in 2005 to 4.34, 63.97, and 31.69 in 2010, respectively. The proportion of the primary industry will decline, the secondary industry will have a slight drop on the basis of relative stability, and the tertiary industry is showing an obvious rising trend. The industrial structure of the three major industries will become more and more reasonable.

The rapid development of the secondary industry and tertiary industry of M City will inevitably lead to the prompt decline in the proportion of the primary industry. Due to the own characteristics of M City, the secondary industry dominates absolutely in the economic development, which is different from the national average and developed countries in the world, and its share cannot be dropped largely in a short period of time. The tertiary industry of M City needs a process to grow and prosper because of its weak basis and slow development. The increase, even a small rise in the proportion of the tertiary industry of M City, needs its growth rate higher than the average economic growth rate, which should be a severe test to the current development of the tertiary industry of M City.

Through the industrial structure optimization, the employment structure of the three major industries will show significant changes, being adjusted from 33.42, 38.02, and 28.56 in 2005 to 31.64, 31.21, and 37.15 in 2010, respectively. The employment proportion of the primary industry and secondary industry will decrease, and the tertiary industry will increase significantly.

With economic development and expeditious urbanization of M City, the agricultural population will transfer to the city. Although the decrease in the range of employment proportion in the primary industry is larger than the drop in the proportion of income, the comparative labor productivity of the primary industry will continue to decline, and the income difference between urban and rural areas is likely to increase further. The advancement in the technological level and productivity of the secondary industry will make a part of labor force transfer from this industry, which will increase a lot of stress to the employment in cities and towns. The comparative labor productivity of the secondary industry will rise to some extent. The tertiary industry will become an important channel for increasing the employment, the range of increase in the employment proportion of the tertiary industry is smaller than the rate of increase in the proportion of income, and the comparative labor productivity of the tertiary industry remains basically in an average level.

9.4.2 Internal Structure Optimization Program of the Primary Industry in M City

Through the internal structure analysis of the primary industry, we divide the farming industry into four sectors: food, cash crops, fruits and vegetables, and other agricultural products; divide animal husbandry into three sectors: domestic animal husbandry, poultry and other animal products. Consequently, the primary industry is divided into nine sectors, which are food crops, cash crops, fruits and vegetables, other agriculture, forestry, domestic animal, poultry, other animal products, and fisheries. The total social production value of the nine sectors from 1999 to 2003 is shown in Table 9.21.

We can get the following regression equations through further study in the relationship of sectors within the primary industry.

1. The regression equation of production value among food crops, cash crops, fruits and vegetables and other agriculture is given by

$$LS = -0.876QZ - 0.477SC + 3.695QN + 36270.787$$

where LS is the production value of food crops; QZ is the production value of cash crops; SC is the production value of

Table 9.21 Total Social Production Value of the Primary Industry in M City (10,000 RMB)

	1999	2000	2001	2002	2003
Food	45,325	42,861	40,397	41,429	34,193
Cash crops	13,116	16,604.5	20,093	17,618	25,045
Vegetables and fruits	13,538	13,446.5	13,355	15,322	17,897
Other agricultural products	7,278	7,408	7,538	7,613	7,705
Forestry	2,015	2,013	2,011	2,259	2,799
Domestic animal	20,634	19,932.5	19,231	19,401	18,250
Poultry	7,712	7,957.5	8,203	8,353	7,985
Other animal products	3,930	3,798	3,666	4,049	3,274
Fisheries	75,356	74,601.5	73,847	77,882	82,620
Total	188,904	188,623	188,341	193,626	199,768

Source: Statistical Yearbook of M City (2000–2004).

vegetables and fruits; and QN is the production value of other agriculture.

2. The regression equation of production value among food crops and three indexes of animal husbandry is given by

$$\mathrm{LS} = 10.61\mathrm{XL} + 11.191\mathrm{QL} - 16.896\mathrm{QD} - 193478$$

where XL is the production value of domestic animal; QL is the production value of poultry; and QD is the production value of other animals.

3. The regression equation of production value among fisheries and three indexes of animal husbandry is given by

$$\mathrm{YE} = -0.921\mathrm{XL} - 1.653\mathrm{QL} + 4.456\mathrm{QD} + 28041.954$$

where YE is the production value of fisheries.

4. The regression equation of production value among fisheries and cash crops, fruits and vegetables and other agriculture is given by

$$\mathrm{YE} = -0.094\mathrm{QZ} + 2.432\mathrm{SC} - 9.973\mathrm{QN} + 118187.6$$

Taking the regression equations mentioned above and the range of the variables as restrictions, we set up the following grey dynamic linear programming model

$$\max Q_{2004} = \mathrm{LS} + \mathrm{QZ} + \mathrm{SC} + \mathrm{QN} + \mathrm{LY} + \mathrm{XL} + \mathrm{QL} + \mathrm{QD} + \mathrm{YE}$$

$$
\text{s.t.}\begin{cases}
LS = -0.876QZ - 0.477SC + 3.695QN + 36270.787 \\
LS = 10.61XL + 11.191QL - 16.896QD - 193478 \\
LY = -0.921XL - 1.653QL + 1.456QD + 28041.954 \\
YE = -0.094QZ + 2.432SC - 9.973QN + 118187.6 \\
YE = -13.893XL - 27.846QL + 27.743QD + 467681.6 \\
LS + QZ + SC + QN \le 89448 \\
XL + QL + QD \le 37394 \\
30123 \le LS \le 34193 \\
27485 \le QZ \le 29639 \\
20044 \le SC \le 21888 \\
8102 \le QN \le 8842 \\
3035 \le LY \le 3218 \\
17702 \le XL \le 18250 \\
8156 \le QL \le 8623 \\
3460 \le QD \le 4214 \\
87577 \le YE \le 92534
\end{cases} \tag{9.1}
$$

We can get the optimal solution by calculating this grey dynamic linear programming model,

$$Q^*_{2004} = 208688.1$$

$$LS = 30123.26, \quad QZ = 29368.18, \quad SC = 21838.57$$

$$QN = 8118, \quad LY = 3035, \quad XL = 18219.73$$

$$QL = 8156, \quad QD = 3609.37, \quad YE = 87577$$

According to the above solution, we optimize the restriction of eleven variable ranges of the grey dynamic linear programming model

Equation 9.1 appropriately, and keep the food production relatively stable, and then the optimal solution from 2005 to 2010 is available:

$$Q^*_{2005} = 218099$$

$$\text{LS} = 28286,\quad \text{QZ} = 29885,\quad \text{SC} = 23917$$

$$\text{QN} = 8012,\quad \text{LY} = 3232,\quad \text{XL} = 18916$$

$$\text{QL} = 8093,\quad \text{QD} = 4114,\quad \text{YE} = 93644$$

$$Q^*_{2006} = 228696$$

$$\text{LS} = 28004,\quad \text{QZ} = 31297,\quad \text{SC} = 26687$$

$$\text{QN} = 8357,\quad \text{LY} = 3533,\quad \text{XL} = 18212$$

$$\text{QL} = 8323,\quad \text{QD} = 4136,\quad \text{YE} = 100146$$

$$Q^*_{2007} = 239251$$

$$\text{LS} = 27232,\quad \text{QZ} = 32752,\quad \text{SC} = 29289$$

$$\text{QN} = 8297,\quad \text{LY} = 3832,\quad \text{XL} = 18350$$

$$\text{QL} = 8321,\quad \text{QD} = 4267,\quad \text{YE} = 106912$$

$$Q^*_{2009} = 261102$$

$$\text{LS} = 23676,\quad \text{QZ} = 35292,\quad \text{SC} = 35902$$

$$\text{QN} = 8208,\quad \text{LY} = 4343,\quad \text{XL} = 18424$$

$$\text{QL} = 8126,\quad \text{QD} = 4523,\quad \text{YE} = 122608$$

$$Q^*_{2010} = 272437$$

$$\text{LS} = 21098,\quad \text{QZ} = 36715,\quad \text{SC} = 39678$$

$$\text{QN} = 8312,\quad \text{LY} = 4618,\quad \text{XL} = 18499$$

$$\text{QL} = 8211,\quad \text{QD} = 4647,\quad \text{YE} = 130659$$

We can obtain the optimal industrial structure program of each year by calculating the above grey dynamic linear programming model.

Changes in the internal structure of the primary industry in M City can be clearly reflected in Table 9.22. Generally, the proportion of agriculture in the primary industry is reducing every year, and the ratio of food crops and cash crops is also decreasing annually; the development of forestry and fisheries has a certain potential, the proportion of which in the primary industry is rising gradually; the proportion of animal husbandry in the primary industry is showing a downward trend. The total social production value of agriculture will remain at the growth rate of over 4%. The proportions of agriculture and fisheries are close in 2005, but after the optimization in the period of 11th Five-Year Plan, the fisheries will be the largest sector in the primary industry, which is decided by the resource situation in M City that it has little cultivated land and abundant fresh water.

9.4.3 Internal Structure Optimization Program of the Secondary Industry in M City

The increment of the secondary industry of M City accounts for a relatively large share in GDP; it is the secondary industry that supports the economic growth and social development of M City. During the period of 10th Five-Year Plan, the secondary industry of M City has made considerable progress, with the fast growing traditional industries such as iron and steel, M City realized the economic growth and industrial structure optimization, which is driven by speeding up the rapid growth of the processing industry through the development of "industrial park." However, there are some problems in the secondary industry of M City: excessive single internal structure, too high proportion of the raw materials industry, short industrial chain, and low added value, and so on. The internal structure of the secondary industry of M City is yet to be upgraded.

In order to seek the optimal layout and the best structure, we work out the internal structural optimization program of the secondary industry of M City on the basis of grey dynamic linear programming model established in accordance with the flow of the production factors in all industries.

Table 9.22 Structure Optimization Program of the Total Social Production Value within the Primary Industry of M City

	2005	2006	2007	2008	2009	2010
Proportion of agriculture in the primary industry	41.31	41.25	40.78	40.14	39.48	38.83
Proportion of forestry in the primary industry	1.48	1.54	1.60	1.65	1.66	1.69
Proportion of animal husbandry in the primary industry	14.27	13.41	12.93	12.65	11.90	11.51
Proportion of fisheries in the primary industry	42.94	43.79	44.68	45.56	46.96	47.96
Ratio of food crops and cash crops	48.62/51.38	47.22/52.78	45.39/54.61	42.51/57.49	40.15/59.85	36.49/63.51
Production value (10,000 RMB)	218099.4	228696.3	239470.2	252779.8	261101.8	272437.5
Growth speed of production value		4.85	4.62	4.23	4.71	4.34

Through in-depth analysis of the internal structure of the secondary industry, we divide the secondary industry into 12 sectors: mining industry, food and tobacco processing industry, textile clothing and leather industry, wood processing, paper-making industry and the cultural and educational supplies, petroleum refining and chemical, pharmaceutical industry, nonmetallic mineral products industry, metal smelting and fabricated metal products industry, machinery industry, transportation equipment manufacturing, electrical and electronic equipment manufacturing industry, public utilities, construction industry.

Suppose x_1, x_2, …, x_{12} are respectively used to express the GDP (10,000 RMB) of mining industry, food and tobacco processing industry, textile clothing and leather industry, wood processing, paper-making industry and the cultural and educational supplies, petroleum refining and chemical, pharmaceutical industry, nonmetallic mineral products industry, metal smelting and fabricated metal products industry, machinery industry, transportation equipment manufacturing, electrical and electronic equipment manufacturing industry, public utilities, construction industry.

b_1 is the labor input (people); b_2 is the possession of fixed assets (10,000 RMB); b_3 is the energy consumption (tons of standard coal); and b_4 is the demand for electricity (10,000 degrees). a_{11}, a_{12}, …, $a_{1,12}$ are respectively used to express the amount of the labor required for increment of 10,000 RMB of mining industry, food and tobacco processing industry, textile clothing and leather industry, wood processing, paper-making industry and the cultural and educational supplies, petroleum refining and chemical, pharmaceutical industry, nonmetallic mineral products industry, metal smelting and fabricated metal products industry, machinery industry, transportation equipment manufacturing, electrical, electronic and equipment manufacturing industry, public utilities, construction industry (people/10,000 RMB). a_{21}, a_{22}, …, $a_{2,12}$ are respectively used to express the fixed assets tied up for increment of 10,000 RMB of mining industry, food and tobacco processing industry, textile clothing and leather industry, wood processing, paper-making industry and the cultural and educational supplies, petroleum refining and chemical, pharmaceutical industry, nonmetallic mineral products industry, metal smelting and fabricated metal products industry, machinery industry, transportation equipment manufacturing, electrical and electronic equipment manufacturing industry, public utilities, construction industry (10,000 RMB/10,000 RMB). $a_{31}, a_{32}, \ldots, a_{3,12}$ are respectively used to express the energy consumption required for increment of 10,000 RMB of mining industry, food and tobacco processing industry, textile clothing and leather industry, wood processing, paper-making industry and the cultural and educational supplies, petroleum refining and chemical, pharmaceutical industry, nonmetallic mineral products industry, metal smelting and fabricated metal products industry, machinery industry, transportation equipment manufacturing, electrical and electronic equipment manufacturing industry, public utilities (tons of standard coal/10,000 RMB) (the energy consumption of construction industry is not taken into consideration because of the lack of the relevant data). $a_{41}, a_{42}, \ldots, a_{4,12}$ are respectively used to express the electrical energy required for increment of 10,000 RMB of mining industry, food and tobacco processing industry, textile clothing and leather industry, wood processing, paper-making industry and the cultural and educational supplies, petroleum refining and chemical, pharmaceutical industry, nonmetallic mineral products industry, metal smelting and fabricated metal products industry, machinery industry, transportation equipment manufacturing, electrical and electronic equipment manufacturing industry, public utilities, construction industry (10,000 degrees/10,000 RMB);

$$\max Z = x_1 + x_2 + x_3 + \cdots + x_{12}$$

$$\begin{cases} a_{11}x_1 + a_{12}x_2 + \cdots + a_{1,12}x_{12} \leq b_1 \\ a_{21}x_1 + a_{22}x_2 + \cdots + a_{2,12}x_{12} \leq b_2 \\ a_{31}x_1 + a_{32}x_2 + \cdots + a_{3,12}x_{12} \leq b_3 \\ a_{41}x_1 + a_{42}x_2 + \cdots + a_{4,12}x_{12} \leq b_4 \\ x_1, x_2, x_3, \ldots, x_{12} \geq 0 \end{cases}$$

For b_1, b_2, b_3, b_4, we construct GM(1,1) model respectively based on the data from 1999 to 2003, to predict their constraint volume of each year from 2006 to 2010.

For $a_{11}, a_{12}, \ldots, a_{1,12}, a_{21}, a_{22}, \ldots, a_{2,12}, a_{31}, a_{32}, \ldots, a_{3,11}, a_{41}, a_{42}, \ldots, a_{4,12}$, we also construct GM(1,1) model respectively in accordance with the data from 1999 to 2003 to predict the resource consuming coefficient of each year from 2006 to 2010. Then we can construct the grey dynamic linear programming model of the secondary industry for each year from 2006 to 2010, to obtain the programming scheme of internal industrial structure and employment structure of the secondary industry during the period of the 11th Five-Year plan.

The optimization program of GDP structure within the secondary industry of M City during the period of 11th Five-Year Plan is shown in Table 9.23.

We find from the Table 9.23 that food and tobacco processing industry, wood processing, paper-making industry and the cultural and educational supplies, transportation equipment manufacturing and construction industry are showing an upward trend in the proportion. The proportion of mining industry, textile clothing and leather industry, petroleum refining and chemical, pharmaceutical industry, nonmetallic mineral products industry, metal smelting and fabricated metal products industry, machinery industry, electrical and electronic equipment manufacturing industry and public utilities shows a downward trend.

The optimization program of employment structure within the secondary industry from 2005 to 2010 can be gained from the

Table 9.23 Optimization Program of GDP Structure within the Secondary Industry in M City (%)

INDUSTRY	2005	2006	2007	2008	2009	2010
Mining	5.60	5.29	5.04	4.82	4.59	4.37
Food and tobacco processing	1.33	1.40	1.43	1.46	1.49	1.52
Textile clothing and leather	1.17	1.10	1.04	0.99	0.94	0.90
Wood processing, paper-making industry and the cultural and cducational supplies	1.92	2.15	2.29	2.40	2.51	2.63
Petroleum refining, chemical, pharmaceutical	1.19	1.16	1.14	1.13	1.11	1.09
Nonmetallic mineral products	0.321	0.322	0.322	0.321	0.318	0.316
Metal smelting and fabricated metal products	61.70	61.35	60.88	60.11	59.64	59.15
Machinery	1.28	1.26	1.25	1.25	1.24	1.23
Transportation equipment manufacturing	1.77	1.94	2.12	2.31	2.43	2.54
Electrical and electronic equipment manufacturing	0.293	0.291	0.289	0.288	0.286	0.283
Public utilities	9.13	8.99	8.92	8.89	8.82	8.75
Construction	14.26	14.68	15.23	15.99	16.58	17.19

prediction of the labor force needed for increment of 10,000 RMB and GDP prediction of each sector within the secondary industry (concrete data are omitted).

We find from Table 9.24 that the majority of the industries are showing a downward trend in the proportion, with the exception of the textile clothing and leather industry, transportation equipment manufacturing sector, whose proportions are rising to a certain extent.

9.4.4 Internal Structure Optimization Program of the Tertiary Industry in M City

During the period of the 10th Five-Year Plan, the growth rate of tertiary industry of M City is obviously too slow, which is not conducive to economic growth and optimization of industrial structure. M City should make use of its own advantage to guide the development of the tertiary industry and internal structure optimization in promoting economic growth and optimization of industrial structure, and give full play to resolve the employment of the tertiary industry during the period of the 11th Five-Year Plan.

Table 9.24 Optimization Program of Employment Structure within the Secondary Industry in M City (%)

INDUSTRY	2005	2006	2007	2008	2009	2010
Mining	12.58	12.22	11.55	10.89	10.47	9.85
Food and tobacco processing	1.47	1.41	1.26	1.12	1.02	0.92
Textile clothing and leather	5.21	5.49	5.70	5.88	6.14	6.36
Wood processing, paper-making industry and cultural and educational supplies	1.11	1.13	1.06	0.98	0.92	0.84
Petroleum refining and chemical, pharmaceutical	0.99	0.84	0.69	0.56	0.47	0.38
Nonmetallic mineral products	1.05	1.01	0.94	0.86	0.81	0.74
Metal smelting and fabricated metal products	26.24	22.71	22.48	22.14	21.08	20.18
Machinery	2.83	2.57	2.26	1.99	1.79	1.57
Transportation equipment manufacturing	2.79	3.88	5.19	6.77	8.56	10.34
Electrical and electronic equipment manufacturing	0.67	0.63	0.57	0.51	0.47	0.43
Public utilities	3.69	3.57	3.36	3.16	3.03	2.84
Construction	41.34	44.51	44.89	45.09	45.21	45.53

In order to find the optimal program of the internal structure in the tertiary industry in M City, we still study the optimization of industrial structure based on the flow of production factors in the various departments, establishing the grey dynamic linear programming model to work out the optimization program of the internal structure of the tertiary industry in M City.

Through in-house study in the tertiary industry, we divide the tertiary industry into six sectors, which are geological survey and water resources management sector, transportation, warehousing, posts and telecommunications sector, wholesale and retail trade catering sector, finance and insurance, real estate and social services sector, culture, education, health and scientific research sector and other industries.

$x_1, x_2, \ldots, x_6$ are respectively used to express GDP (10,000 RMB) of geological survey and water resources management, transportation, warehousing, posts and telecommunications, wholesale and retail trade catering, finance and insurance, real estate and social services, culture, education, health and scientific research and other industries.

b_1 is the labor input (10,000 people); b_2 is the basic construction investment (10,000 RMB); b_3 is the demand for electricity (10,000 degrees). $a_{11}, a_{12}, \ldots, a_{16}$ are respectively used to express the labor required for the production value of 10,000 RMB of the above industries (people/10,000 RMB). $a_{21}, a_{22}, \ldots, a_{26}$ are respectively used to express the basic construction investment required for the production value of 10,000 RMB of the above industries (10,000 RMB/10,000 RMB). $a_{31}, a_{32}, \ldots, a_{36}$ are respectively used to express electricity consumption required for the production value of 10,000RMB of the above industries (10,000 degree/10,000 RMB).

Then, we establish the grey dynamic linear programming model as follows:

$$\max Z = x_1 + x_2 + x_3 + \cdots + x_6$$

$$\begin{cases} a_{11}x_1 + a_{12}x_2 + \cdots + a_{16}x_6 \leq b_1 \\ a_{21}x_1 + a_{22}x_2 + \cdots + a_{26}x_6 \leq b_2 \\ a_{31}x_1 + a_{32}x_2 + \cdots + a_{36}x_6 \leq b_3 \\ \\ x_1, x_2, x_3, \ldots, x_6 \geq 0 \end{cases}$$

Table 9.25 Optimization Program of the Internal Increment Structure in the Tertiary Industry in M City (%)

	2005	2006	2007	2008	2009	2010
Geological survey and water resources management	1.45	1.40	1.35	1.30	1.25	1.21
Transportation, warehousing, posts and telecommunications	20.24	19.69	19.14	18.61	18.08	17.56
Wholesale and retail trade catering	23.83	24.23	24.62	25.00	25.39	25.78
Finance and insurance, real estate and social services	37.55	37.83	38.11	38.38	38.64	38.88
Culture, education, health and scientific research	12.32	12.41	12.50	12.59	12.67	12.75
Other industries	4.61	4.44	4.28	4.12	3.97	3.82

For b_1, b_2, b_3, b_4, we construct GM (1,1) model respectively according to the data of the year from 1999 to 2003 to predict the various constraint volumes of the year from 2006 to 2010. For $a_{11}, a_{12}, \ldots, a_{16}, a_{21}, a_{22}, \ldots, a_{26}, a_{31}, a_{32}, \ldots, a_{36}$, we also construct GM (1,1) model respectively according to the data of the year from 1999 to 2003 to predict all kinds of resource consuming coefficients.

Then we can construct the grey dynamic linear programming of each year from 2006 to 2010 respectively to get the optimal program of the internal industrial structure of the tertiary industry of M City during the period of 11th Five-Year Plan.

We find from Table 9.25 that the proportions of wholesale and retail trade catering, finance and insurance, real estate and social services, Culture, education, health and scientific research in the tertiary

Table 9.26 Optimization Program of Labor Force Structure within the Tertiary Industry of M City (%)

	2005	2006	2007	2008	2009	2010
Geological survey and water resources management	0.878	0.77	0.67	0.59	0.52	0.45
Transportation, warehousing, posts and telecommunications	16.65	16.59	16.50	15.64	14.66	14.79
Wholesale and retail trade catering	49.78	50.51	51.18	52.26	53.37	53.75
Finance and insurance, real estate and social services	15.52	15.86	16.19	16.67	17.15	17.41
Culture, education, health and scientific research	11.36	11.22	11.06	10.99	10.93	10.71
Other industries	5.80	5.04	4.38	3.83	3.36	2.89

industry are showing an upward trend to a certain extent, while other industries are declining more or less.

The optimization program of labor force structure within the tertiary industry of the year from 2005 to 2010 can be obtained from the prediction of the labor needed for increment of 10,000 RMB and GDP prediction of each sector within the tertiary industry.

We find from Table 9.26 that the employment proportions of wholesale and retail trade catering, finance and insurance, real estate and social services in the tertiary are increasing, while the proportions of other industries are showing a downward trend.

10

Keystones and Ideas of the Industrial Structure Adjustment in S Province

S Province, located in the coastal region, has been experiencing unceasing improvement in productive force and moving toward a high grade level of industrial structure step by step, after more than two decades of reform, opening up and development. However, in the 21st century, with the formation of a new pattern of economic globalization and China's entrance to the WTO, S Province is facing a very serious and urgent structure adjustment, and the contradictions of maladjustment in the market economy mechanism under the new situation of economic structure have become increasingly prominent. Especially, the optimization and upgrading of industrial structure is a very urgent strategic task. It is of particular significance for S Province to carry out the industrial structure adjustment to improve the overall quality of its economy, enhance its competitiveness in the international market, and maintain a healthy and rapid economic development. As a result, industrial structure adjustment, industrial structure upgrading, selection of leading industries, and exertion in regional comparative advantages in speeding up economic development have become a major task of S Province.

10.1 Current Situation and Problems in the Industrial Structure in S Province

10.1.1 Current Situation Analysis of the Industrial Structure in S Province

10.1.1.1 Comparative Analysis of the Industrial Structure between S Province and Other Provinces in the Country As S Province is located in the eastern coastal area, many relevant factors lead to certain differences in the industrial structure between S Province and other provinces.

Table 10.1 Comparison of the Industrial Structure with Other Provinces (Municipalities) in China

	GDP STRUCTURE IN THE INDUSTRY (%)			EMPLOYMENT STRUCTURE IN THE INDUSTRY (%)		
REGION	PRIMARY	SECONDARY	TERTIARY	PRIMARY	SECONDARY	TERTIARY
China	15.9	50.9	33.2	50.0	22.5	27.5
S Province	12.0	51.7	36.3	42.7	30.2	27.1
Zhejiang Province	11.0	52.7	36.3	37.8	30.9	31.3
Shanghai city	1.8	47.5	50.6	13.1	42.8	44.1
Guangdong Province	10.4	50.4	39.3	41.1	26.2	32.7
Shandong Province	14.9	49.7	35.5	53.1	23.6	23.3

From Table 10.1 we find that there is little difference between S Province and provinces of Guangdong, Zhejiang, and Shandong for GDP proportion of three major industries; the proportion of increment in GDP of the primary industries is about 11–15%, that of the secondary industry is about 50%, and that of the tertiary industry is about 35%. The proportion of the tertiary industry in Shanghai is obviously higher than those of the other four provinces, while the proportion of the primary industry in Shanghai is much lower, which reflects the characteristics of Shanghai's economy.

10.1.1.2 Analysis of the Internal Structure of the Primary Industry in S Province With the economic development, a significant change has happened in the industrial structure of the primary industry of S Province. The proportion of farming has declined, while the proportions of animal husbandry and fisheries in particular have increased. It can be seen from Table 10.2 that the internal industrial structure of the primary industry remained basically unchanged before 1978, which is largely determined by industrial policies of that time. After the reform and opening up to the outside world, great changes have taken place in the industrial structure. Compared with that in 1978, the proportion of farming has fallen by 21.83%; the proportion of fisheries has increased by 14.44%; that of animal husbandry has had a growth of 7.18%.

Table 10.2 Changes in the Structure of total Production Value in Agriculture, Forestry, Animal Husbandry and Fisheries in S Province (%)

YEAR	1952	1957	1962	1978	1980	1985	1990	1995	2000
Farming	82.02	81.77	85.16	80.45	76.55	69.55	62.44	58.46	58.62
Forestry	0.09	0.60	0.80	1.40	1.40	1.60	1.37	1.27	1.61
Animal husbandry	15.69	14.26	11.16	15.85	19.25	23.06	27.70	28.20	23.03
Fisheries	2.20	3.37	2.89	2.30	2.80	5.38	8.50	12.07	16.74

If agriculture is divided into two departments, farming (mainly in food production) and other sectors, we find from a horizontal comparison that with the growth of per capita GDP, the proportion of farming in total production value of agriculture is gradually declining, while that of other sectors is increasing.

S Province is a major agricultural province; the agricultural industrial structure adjustment will have a direct effect on the entire economic development and structural adjustment of S Province. It can be found from the internal structure of agriculture in both China and S Province that the proportion of faming in agriculture is declining gradually, while those of other sectors are increasing. In the internal structure of agriculture, the proportion of farming is 3% higher than the national average level, and about 10% higher than those of Zhejiang, Guangdong (see Table 10.3), which shows that the adjustment of agricultural structure in S Province lags behind the national average level, and the difference is larger compared with other provinces at the similar economic level; therefore, there will be even more difficult task on the agricultural industrial structure adjustment of S Province.

Table 10.3 Comparison of Internal Industrial Structure in Agriculture in the Country and Some Provinces (Municipalities) in 2000 (%)

REGION	FARMING	OTHER SECTORS
China	55.68	44.32
S Province	58.62	41.38
Zhejiang Province	48.99	51.01
Shanghai city	41.48	58.52
Guangdong Province	49.24	50.76
Shandong Province	56.70	43.30

10.1.1.3 Analysis of the Internal Structure of the Secondary Industry in S Province With the rapid development of economy in S Province, great changes have taken place in the industrial structure of the secondary industry, and the most prominent change is the rapid increase in the proportion of heavy industries, which rose from 6.1% in 1952 to 47.6% in 1978, an increase of 41.5%, and then the proportion has been fluctuating around this level.

It can be seen from Table 10.4 that industries such as machinery, chemicals, food, and textile have large changes in the proportion of the industrial output values in S Province. Their main features are as follows: First, the proportion of output of machinery industry increased every year and became the largest industry of S Province; the ratio rose from 2% in 1952 to 29.7% in 1985 and kept stable essentially for several years. Before 1985, machinery industry developed rapidly because there has been a great demand for it due to the rapid development of heavy industry. After 1985, although the ratio has been kept relatively stable, the machine-building industry still had a large internal adjustment, which has been mainly represented by the continuous declining demand in agricultural machinery and industrial machinery, rapid growth in machinery for daily use and unceasing changes in consumer hotspots. Second, the proportion of output of the chemical industry was on the rise, from 0.4% in 1952 to 15% in 1998, increasing by 14.6%. Third, the output value of food industry

Table 10.4 Changes in the Industrial Structure of the Secondary Industry of S Province (%)

INDUSTRY	1952	1957	1965	1978	1985	1990	1995	1998
Light	93.89	84.81	75.25	52.4	53.3	54.7	48.8	51.9
Heavy	6.11	15.19	24.75	47.6	47.7	45.3	51.2	48.1
Metallurgical	0.3	0.5	2.5	3.4	4.8	5.66	6.5	4.6
Electric power	1.0	1.1	2.2	3.0	2.0	2.14	3.2	7.6
Coal	1.3	1.7	1.6	1.4	1.0	0.8	0.6	1.3
Oil	–	0.7	3.2	2.6	1.6	2.1	1.7	–
Chemical	0.4	1.6	6.3	13.4	12.2	15.4	13.6	15.0
Machinery	2.0	5.8	11.7	25.7	29.7	23.3	28.4	29.5
Building materials	2.2	3.5	5.0	5.9	6.1	5.0	5.3	4.1
Food	40.6	37.8	26.1	10.8	10.1	8.2	6.5	7.0
Textiles	40.1	36.0	33.5	23.5	22.3	25.5	19.0	13.9
Paper	0.7	1.5	1.0	2.7	0.7	1.2	1.0	1.1

dropped largely in proportion, which made the food industry become the fifth largest industry in 1998 from the largest industry in 1952; for the internal structure of the food industry, the proportion of beverage and other food increased faster on account of the advance of the food consumption level. Fourth, the textile industry became the third largest industry in 1998 from the second largest industry in 1952, and the proportion declined from 40.1% in 1952 to 13.9% in 1998, dropping nearly by 26%.

10.1.1.4 Analysis of the Internal Structure of the Tertiary Industry in S Province Since 1978, the tertiary industry has developed rapidly in S Province with an average annual growth rate of 15%, which is above the average GDP growth rate in the same period. In 2000, the total proportion of catering business, transportation, posts and telecommunications, finance and insurance, real estate in the tertiary industry is more than 80% (see Table 10.5).

The internal structure of the tertiary industry in S Province has the following characteristics. First, the proportion of catering business in the tertiary industry fluctuated during the period, but it was always at the top of the tertiary industry; since the reform and opening up to the outside world, business in the catering industry has kept a good development momentum, and has become one of the important symbols of the prosperity and development in the tertiary industry. Second, transportation, posts and telecommunications industry developed rapidly; its economic foundation has been strengthened and the "bottlenecks" of development has been eased significantly, but the proportion in the tertiary industry did not rise too fast, its growth rate

Table 10.5 Changes in the Internal Structure of the Tertiary Industry in S Province (%)

INDUSTRY	1978	1980	1985	1990	1995	2000
Tertiary	100.0	100.0	100.0	100.0	100.0	100.0
Transportation, posts and telecommunications	13.9	13.9	22.4	17.3	15.9	17.9
Catering business	27.6	31.0	28.2	28.0	32.5	27.5
Finance and insurance	21.1	19.8	16.2	19.7	17.0	13.8
Real estate	11.1	10.5	10.2	7.3	9.0	11.3
Social services	5.9	5.5	4.7	5.4	6.5	8.5
Others	20.4	19.3	18.3	22.3	19.1	21.0

Table 10.6 Changes in the Internal Structure of the Tertiary Industry of Other Provinces in China (%)

	TRANSPORTATION, POSTS AND TELECOMMUNICATIONS		CATERING BUSINESS		FINANCE AND INSURANCE		REAL ESTATE		OTHERS	
PROVINCE	1995	2000	1995	2000	1995	2000	1995	2000	1995	2000
Shanghai	17.2	13.7	27.1	21.1	24.9	29.7	9.2	10.9	21.7	24.6
Zhejiang	18.1	19.6	41.7	37.8	12.5	9.6	5.0	5.3	22.7	27.7
Guangdong	20.6	23.9	23.1	25.5	18.1	9.8	8.1	13.6	30.2	27.2
Shandong	18.5	18.3	27.2	25.9	17.6	14.9	9.9	10.2	26.9	30.7
S Province	15.9	17.9	32.5	27.5	17.0	13.8	9.0	11.3	25.6	29.5

is slightly higher than the average speed of the tertiary industry. Third, the proportion of the increment of finance and insurance industry had a downward trend every year, falling from 21.1% in 1978 to 13.8% in 2000, with a decrease of 7.3%. Fourth, real estate is a newly emerging industry in the tertiary industry, which has become a new economic growth point of the tertiary industry in S Province due to its rapid development and its strong correlation with other industries.

There are some differences in the internal structure of the tertiary industry between S Province, Shanghai, Zhejiang Province, Guangdong Province and Shandong Province (Table 10.6). Especially in Shanghai, the proportion of its finance and insurance industry in the tertiary industry is rising, which reflects the status of Shanghai as China's financial center which has been gradually strengthened.

10.1.2 Characteristics of the Evolution of the Industrial Structure in S Province

The evolution of the industrial structure in S Province is characterized by its imbalance. According to the "typical structure" analysis of Chenery, in terms of static state, the current state of the industrial structure in S Province is at a low-income status, and the income proportion of the secondary industry is too high. With respect to dynamic resource conversion, before the 1980s, it was difficult to make potential demand into real demand due to low per capita income of residents, the final demand only had a very small impact on the evolution of the industrial structure; allocation of resources inclined to the secondary industry, especially to the heavy industry, the industrial structure evolved along the track of the rapid increase in the proportion of industry, soon leading to disequilibrium in the industrial structure with an excessively high proportion in the secondary industry. After the 1980s, with the rapid increase in per capita income, potential demand has gradually becoming real demand, consumption structure has changed considerably, and the demand exerts a larger impact on the evolution of the industrial structure, which expedites the balanced evolution of the industrial structure. In addition, before the 1980s, the allocation of resources mainly inclined to heavy industry, not to light processing industry that could meet the consumers` demand for daily requirements, resulting in too small a proportion of light processing industry. After the 1980s, light processing industry is showing a rapid momentum of development pushed by the demand. However, heavy processing industry is restricted by the irreversible resource allocation, so it is difficult to transfer its stock resources into other sectors; although the industrial structure as a whole tends to a balanced evolution, the actual process is still at the state of disequilibrium.

The slow development of the tertiary industry is mainly reflected in the disequilibrium in the industrial structure of S Province. Before the 1980s, in the process of resources inclining to heavy industry, there was no strong correlation among industries; the industrial development had a strong self-cycle. At the current situation, when we evaluate the tertiary industry whether from the proportion of employment or from the proportion of income, it has a lower status in the structure; as for the internal structure of the tertiary industry, the traditional flow of business sectors still account for the major proportion, while sectors on behalf of the modern high grade industrial direction such as scientific research, integrated technology services, consulting services, and finance and insurance are still at a very low level of development. Transportation, posts and telecommunications and other industries, which should have an advanced development in the process of industrialization, are lagging behind obviously.

It can be found from the evolution trends in the industrial structure that the evolution of the industrial structure in S Province is tending to a favorable circle over time. After the 1980s, the industrial structure has evolved into a benign state. With the implementation of the reform and opening-up policy, as market mechanisms have been introduced into the adjustment mechanism of resources allocation, the effect of the system reform has emerged. As increase in people's income is large, advance in consumer demand level and unceasing changes in consumption structure level have exaggerated the effect of the demand in pulling the evolution of the industrial structure gradually. As a result, the structural efficiency is being improved continuously, and the conversion rate of resources among the three major industries is also being increased gradually.

10.1.3 Problems in the Industrial Structure in S Province

Since the reform and opening up to the outside world, S Province has made great achievements in economic development. Although certain improvement and optimization have been realized in the industrial structure adjustment, there are still some problems that cannot be ignored. They are as follows. The first problem is the low level in the industrial structure. Low level in the industrial structure of S Province is represented not only by the low proportion of the tertiary industry in the three major industries, but also by low proportion of the scientific and technological industries in the three major industries.

10.1.3.1 Low Proportion of the Tertiary Industry in the Three Major Industries According to Simon Kuznets' theory on law of evolution in the industrial structure of the three major industries, in general, the larger the increment of the tertiary industry, the higher the industrial structure level, and vice versa. With a per capita GDP of 11,773 RMB in 2000, S Province was at the mid-industrialization level. In terms of industrialization, the tertiary industry in S Province was at a lower level (Table 10.7).

In respect of industrial structure, there is a little difference in the GDP proportion of the three major industries between S Province, Zhejiang Province, Guangdong Province, and Shandong Province. The increment of the primary industry accounted for 10–15%, the proportion of the secondary industry is 50–53%, and that of the tertiary industry is about 35%. The proportion of the tertiary industry of Shanghai is more significant than that in other four provinces,

Table 10.7 Comparison of GDP Structure of the Three Major Industries (%)

REGION	YEAR	PRIMARY	SECONDARY	TERTIARY
S Province	2000	12.0	51.7	36.3
Zhejiang Province	2000	11.0	52.7	36.3
Shanghai Province	2000	1.8	47.5	50.6
Guangdong Province	2000	10.4	50.4	39.3
Shandong Province	2000	14.9	49.7	35.5
China	2000	15.9	50.9	33.2
USA	1998	1.7	26.2	72.1
Japan	1998	1.7	37.2	61.1
France	1998	2.3	26.2	71.5
England	1998	1.8	31.5	66.7
India	1998	29.3	24.7	45.9

showing the characteristics of its urban economy. As to the production value of the three major industries, the secondary industry ranks no. 1, the tertiary industry no. 2, and the primary industry no. 3 in S Province (that is, "the secondary industry, the tertiary industry, the primary industry" or "two, three and one" in order of the production value); While in India, the feature is "three, one, two"; that is to say, the industrial structure level of S Province is lower than that of India. Compared with developed countries, there is still a long way to go in the industrial restructuring in S Province.

10.1.3.2 Low Proportion of Scientific and Technological Industries in the Three Major Industries Within the three major industries, the proportion of high-tech products was not large, the R&D funds was too low, which was about 5670 million RMB in 2000, accounting for 0.66% of GDP, and the investment structure is imbalanced. High-tech industrial output value reached 177,550 million RMB, taking up 10.1% of the total industrial output value of the province, while the high-tech industrial output value of Shanghai in 1997 was 15% in the total industrial output value. Technological innovation has not yet become an independent action of enterprises, which are still taking the old extension-style road with expansion of reproduction.

The second problem is the nonconformity in the industrial structure and labor structure. As part of the industrial structure, the employment structure should have consistent development and changes with the industrial structure. Accompanied by economic development, the proportion of labor force in the primary industry is reduced gradually, while the proportion of the secondary industry and the proportion of the tertiary industries will be increased gradually.

S Province had significant changes in the employment structure, the ratio of employees in the three major industries became 40.7%, 30.2%, and 27.1% in 2000 from 69.7%, 19.6%, and 10.7% in 1978; the employment proportion of the primary industry declined by 29%, while the proportion of the secondary and the tertiary industry significantly increased (shown in Table 10.8). As to the employment structure, the primary industry ranked no. 1, the secondary industry no. 2, and the tertiary industry no. 3 in S Province (that is the feature of "one, two, three" in the order of the employment structure); the employment proportion in the primary industry accounted for 42.7% in the total labor force, which was obviously different from the "two, three, one" situation in terms of the output value structure. The conversion rate of the employment structure significantly lagged behind that of the output value structure in the three major industries. From the horizontal comparison, there were still a certain distance between S Province and advanced provinces in China in the employment structure; for example, Zhejiang, Guangdong had formed the "one, three, two" structure in the order of the employment structure, and the proportion of the secondary industry and tertiary industry nearly equaled the employment structure of Shandong; Shanghai had formed "three,

Table 10.8 Comparison of the Employment Structure in the Three Major Industries (%)

REGION	YEAR	PRIMARY	SECONDARY	TERTIARY
S Province	2000	42.7	30.2	27.1
Zhejiang Province	2000	37.8	30.9	31.3
Shanghai city	2000	13.1	42.8	44.1
Guangdong Province	2000	41.1	26.2	32.7
Shandong Province	2000	53.1	23.6	23.3
China	2000	50.0	22.5	27.5
USA	1998	2.7	23.9	73.4
Japan	1998	5.3	32.0	62.7
UK	1998	1.7	26.6	71.7
Germany	1998	2.9	33.8	63.3
Canada	1998	3.7	22.4	73.9

two, one" state in the order of both the industrial structure and the employment structure. Compared with the developed countries, the employment structure in S Province lagged far behind.

The third problem is the convergence of regional industrial structure over the growing low-level competition. The irrational industrial structure is shown not only in the structure of sectors, but also in the regional industry structure. The convergence of regional structure is quite prominent due to development strategy being in the process of change, coupled with the fragmentation of the management system and the problem of repeated investment and redundant construction in the region, what they have done is to increase local revenue regardless of local resources and industrial structure in S Province. Industries such as machinery, electronics, petrochemical, automobile, construction, and building materials have been listed as supporting industries in the "Ninth Five-Year Plan" and the long-term plan in 2010; while in more than 30 provinces, municipalities and autonomous regions in China, and nearly 26 regions have listed machinery and petrochemical industries as supporting industries, about 25 regions have listed electronics as supporting industry, the number of regions that have listed automobile, construction and building materials as their supporting industries is 22 and 19, respectively, which has fully explained that the industrial structure of S Province is an objective problem. Low-level redundant construction not only leads to the convergence of regional structure, but also causes massive productivity wasted and maks poor economic performances in some areas.

The fourth problem is the short chain of industrial products and the failure in transforming the resource advantages into industrial advantages. Exertion of comparative advantage is one of important principles of industrial structure adjustment. However, S Province could not make good use of this principle in the industrial structure adjustment. Take food industry as an example, S Province is a major agricultural province, but its food industry has not developed well. Food industry is a lasting industry and is extremely important in the national economy. So far, the output value of any industrialized countries in the world in food industry ranks top three places in all the industrial sectors, the output value of USA in food industry is equivalent to the total output of machinery and electronics industry. In 1996, only 11 corporations were engaged in food manufacturing in the world's top 500 enterprises.

S Province is a major agricultural province, and the grain output reached 35,590,300 tons in 1999, ranking fourth in the country. At the same time, S Province is also a major province of food industry. At the end of the period of "Seventh Five-Year Plan," the output value of food industry in S Province once ranked first in the country. However, the rank has become no. 3 now, behind Guangdong and Shandong due to declining investment and other reasons. The proportion of food industry in S Province has been declining every year from 20–33% during the 1950s and 1960s to 10–14% to during the 1970s and 1980s, and declining further to 5–7% in the 1990s. As the product structure could not meet the requirements of the industrialized mass production, there has been an increasingly widening gap in food industry between the S Province and other provinces and municipalities, and it is more and more difficult for S Province to develop food industry resulting in a large number of food products import from other regions.

The fifth is the lack of technological innovation capability and industrial competitiveness. S Province is the first region which put forward the strategy of "developing the province through science and technology," and then adjusted the strategy to "developing province through science and education" in 1994. In recent years, S Province has been gradually speeding up its pace in high-tech enterprises cultivation. By the end of 2000, there were 1223 high-tech enterprises in S Province, with the output value of 177,550 million RMB, taking up 17% of the total output value of the state-owned industrial enterprises and nonstate-owned industrial enterprises with an annual revenue of more than 5 million RMB in the province; the per capita labor productivity of high-tech enterprises is five to ten times higher than that of traditional enterprises. After several years of efforts, S Province has made some progress in technological innovation, but at the same time there are still a number of shortcomings and problems:

1. *Undeveloped product technology, low added value, and weak market competitiveness.* Labor-intensive and primary processing products take up a high rate in the major industrial products

of the province; the proportion of high-tech and high value-added, brand-name products is small. The advantage of the products of S Province is threatened seriously by the rapid industrial development of other brother regions and increasingly fierce market competition where plenty of foreign products enter in a large scale. A great number of new, unique, and excellent products and mechanical and electrical equipment products with high technology and high grade depend on largely on import.

2. *Inadequate combination between science, technology, and economy and lower transformation rate of scientific and technological achievements.* Although there are a large number of colleges, universities and research institutions, the productivity and research are not in closely connected; many scientific and technological achievements drift away from the business; enterprises have a weak ability of technological assimilation or redevelopment and renovations. They often pay more attention to technology that can be applied to the production line directly, but ignore promoting the development of enterprises through technology market, information services and advisory services.
3. *Relative shortage of investment in science and technology.* It is difficult to meet the needs of rapid economic growth despite the investment in technological innovation growing every year in S Province. In 2000, the expenditure on research and development activities in the province accounted for 0.66% in GDP, which is lower than the national average; the expenditure of corporate research and development took up 0.98% in the revenue of industrial products. However, according to internationally acknowledged standards, only when the proportion of corporate research and development funds in the revenue of industrial products reaches more than 8%, can the corporate have strong vitality. Therefore, a great deal of investment in corporate research and development is needed to meet the needs of enterprises in technological innovation.
4. *Nonenhancement of technological innovation of enterprises.* Despite good technical quality in large and medium-sized enterprises,

there are few companies developing in high-tech direction; enterprises have little interest in difficult and long-term research and development projects; high-tech technology relies on introduction from abroad. S Province has failed in making good use of the potential advantages of cooperation among enterprises, colleges, universities, and research institutions.

10.2 Selection of Leading Industries in S Province

As the process of upgrading the industrial structure is represented by changes and development of leading industries, the basic idea of the adjustment and optimization in the regional industrial structure is as follows: Selection and optimization of leading industries can effectively take on the task of national regional division, and enhance its radiation power to drive regional economic development; support the development of related industries, particularly develop the forward related industries of leading industries to extend the product chain as much as possible; improve the industrial quality of the leading industries to maintain and consolidate its role, and actively develop basic industries that needs to be balanced, especially the "bottleneck" industries to overcome its constraints in regional economy; support general industries so that after the position of original leading industries is weakened due to changes of condition, new leading industries can take over the position to maintain the normal metabolism of the regional economic system.

10.2.1 Evaluation Index System in the Selection of Leading Industries in S Province

Combined with the actual situation in S Province, we have selected the following indexes as the index system in the selection of regional leading industries.

1. *Industrial relative index:* There are two commonly used industrial relative indexes, which are indexes of sensitivity factor and influence factor. Suppose b_{ij} is a coefficient of the complete consumption coefficient matrix B in the input–output analysis, n is the number of industrial sectors, then the index

of sensitivity factor of sector i is given by

$$\mu_i = \frac{\sum_{j=1}^{n} b_{ij}}{(1/n)\sum_{i=1}^{n}\sum_{j=1}^{n} b_{ij}}$$

and the index of influence factor of sector j is given by

$$\gamma_i = \frac{\sum_{i=1}^{n} b_{ij}}{(1/n)\sum_{i=1}^{n}\sum_{j=1}^{n} b_{ij}}$$

2. *Income elasticity of demand index:* Suppose y is the GDP, x_i is the demand for products of sector i, then the income elasticity of sector i is

$$\varepsilon_i = \frac{y}{x_i}\frac{\partial x_i}{\partial y}$$

3. *Growth rate index:* Suppose x_i^0 is the demand for products of sector i at the initial state, r_i is the average growth rate, and then the demand for products of sector i is $x_i^t = x_i^0(1+r_i)^t$
4. *Employment index:* We use labor force factor of comprehensive employment to evaluate capacity of each industry to absorb labor force. Suppose v_i is the labor remuneration in inpvut–output table, q_i is the total output, c_{ij} is the coefficient of Leontief inverse matrix, and then the labor force factor of comprehensive employment of sector i is given by $\beta_i = \Sigma(v_i/q_i)c_{ij}$. These indexes are all dimensionless ones. In order to facilitate the evaluation, we will divide the regional department into three different types of industries in accordance with its importance in the regional economic development, which are general industries, auxiliary industries and leading industries, and then determine various types of whitened weight function and the weight of each index in the comprehensive evaluation through the Delphi survey.

10.2.2 Selection of Leading Industries and Auxiliary Industries in S Province

Based on the input–output table of S Province in 1997 and the statistical yearbook of S Province over the years, we undertake the selection of leading industries and auxiliary industries by means of grey fixed clustering model for the selection of leading industries.

1. The sensitivity factor, influence factor and comprehensive employment rate of each sector in 1997 can be calculated from the input–output table.
2. The income elasticity of demand and the growth rate can be calculated from the statistical yearbook over the years.
3. On account of the inconsistency in the sector selection between statistical yearbook basis and input–output table basis, we have merged the 40 sectors in the input–output table into 34 sectors. The results of index value are in Table 10.9.

According to the above grey fixed clustering model and each index value, we suppose f_j^k (*) is the whitened weight function of index j and grey type k, and then determine various types of whitened weight functions and weight of each index in the comprehensive assessment through Delphi survey. According to collection and analysis of the investigated data, whitened weight function of sensitivity factor, influence factor, income elasticity factor, growth rate, labor force factor of comprehensive employment are, respectively, as follows:

$$f_1^1[-,-,0,0.5],\quad f_1^2[0.5,-,1,1.5],\quad f_1^3[0,1.3,-,-]$$

$$f_2^1[-,-,0.5,1.2],\quad f_2^2[0.5,-,0.9,1.3],\quad f_2^3[0.6,1.1,-,-]$$

$$f_3^1[-,-,0.4,1.2],\quad f_3^2[0.6,-,1.3,2.0],\quad f_3^3[1,1.8,-,-]$$

$$f_4^1[-,-,0.07,0.2],\quad f_4^2[0.05,-,0.12,0.19],\quad f_4^3[0.1,0.18,-,-]$$

$$f_5^1[-,-,0.37,0.57],\quad f_5^2[0.3,-,0.5,0.7],\quad f_5^3[0.35,0.6,-,-]$$

The weight functions of sensitivity factor, influence factor, income elasticity factor, growth rate, labor force factor of comprehensive employment are 0.14, 0.14, 0.24, 0.24, and 0.24, respectively. The

Table 10.9 Results of Index Value of 34 Sectors in S Province

INDUSTRY	SENSITIVITY FACTOR, μ_i	INFLUENCE FACTOR, γ_i	INCOME ELASTICITY OF DEMAND, ε_i	GROWTH RATE, r_i	COMPREHENSIVE EMPLOYMENT RATE, β_i
Agriculture	0.9633	0.6756	0.260	0.008	0.7286
Coal selection	0.8889	0.6876	0.526	−0.009	0.5936
Oil and natural gas exploration	0.8199	0.3589	3.111	0.337	0.3528
Metallic mine selection	0.2161	0.7366	0	0.065	0.6717
Nonmetallic mine selection	0.6359	0.7649	2.676	−0.111	0.4226
Food and tobacco processing	0.8249	1.0321	0.791	0.072	0.5096
Textile	1.2055	1.2583	1.256	0.064	0.3982
Clothing leather, plume and fibre products	0.2598	1.1266	1.479	0.090	0.6244
Wood processing and furniture manufacturing	0.2896	1.0386	1.227	0.130	0.4542
Paper-making and printing, culture and education goods	0.9162	1.1368	1.363	0.114	0.4746
Oil processing and coking	1.2970	0.9357	0.759	0.115	0.3630
Chemical	3.3553	1.0795	1.170	0.136	0.3834
Nonmetallic mineral products	1.1903	1.0556	1.757	0.017	0.4596
Metal smelting and rolling processing	4.7383	1.4733	2.084	0.020	0.4607
Fabricated metal products	0.8404	1.3129	0.889	0.110	0.4002
Machinery	1.2090	1.2996	1.199	0.046	0.4296
Transportation equipment manufacturing	0.6622	1.5949	0.766	0.066	0.4370
Electrical machinery and equipment manufacturing	0.8473	1.4695	1.744	0.119	0.4154
Electronics and telecommunications equipment manufacturing	1.1501	1.2383	2.511	0.262	0.3957
Instruments and office machinery manufacturing	0.1256	1.2873	3.172	0.211	0.4362
Other manufacturing	1.1311	1.0108	5.764	0.044	0.5896
Electricity, steam and hot water production and supply	1.6183	0.7644	1.294	0.232	0.4471
Gas production and supply	0.0946	1.4203	1.789	0.142	0.6892
Water production and supply	0.1208	0.7659	0.305	0.042	0.5050
Construction	0.3474	1.1201	1.262	0.147	0.5311

continued

Table 10.9 (continued) Results of Index Value of 34 Sectors in S Province

INDUSTRY	SENSITIVITY FACTOR, μ_i	INFLUENCE FACTOR, γ_i	INCOME ELASTICITY OF DEMAND, ε_i	GROWTH RATE, r_i	COMPREHENSIVE EMPLOYMENT RATE, β_i
Transportation, posts and telecommunications	1.8850	0.6446	0.850	0.078	0.4722
Catering business	3.1201	0.7710	0.867	0.046	0.5482
Finance and insurance	1.1328	0.5166	1.021	0.145	0.2747
Real estate	0.2238	0.9941	1.236	0.169	0.4061
Social services	1.1264	0.7829	1.227	0.143	0.5008
Sanitation and physical education, social welfare	0.2152	1.2420	1.699	0.178	0.5205
Education, arts, radio, film and television	0.1110	0.7258	0.988	0.111	0.7162
Comprehensive scientific research and technical service	0.1969	0.8553	0.902	0.135	0.5796
The executive authorities and other sectors	0.2412	0.9139	1.851	0.227	0.5478

integrated clustering analysis results of 34 sectors in S Province are shown in Table 10.10.

The conclusions obtained through the above calculation are as follows:

First, the leading industry candidates are below oil and natural gas exploration, nonmetallic mine selection industry, clothing, leather, plume and fiber products, nonmetallic mineral products industry, electrical machinery and equipment manufacturing industry, electronics and telecommunications equipment manufacturing, instruments and office machinery manufacturing, other manufacturing, electricity, steam and hot water production and supply, gas production and supply industry, sanitation, physical education and social welfare sector.

Second, the auxiliary industry candidates are the food and tobacco processing industry, textile industry, wood processing and furniture manufacturing, paper-making and printing, culture and education goods industry, chemical industry, fabricated metal products, machinery, construction industry, real estate, social services, education, arts, radio, film and television industry, comprehensive scientific research and technical service.

Third, the general industry candidates are agriculture, coal selection industry, metallic mine selection industry, oil processing and

Table 10.10 Integrated Clustering Evaluation Results of 34 Sectors in S Province

NO.	INDUSTRY	GENERAL INDUSTRY	AUXILIARY INDUSTRY	LEADING INDUSTRY	MAX	CLUSTERING RESULTS INDUSTRY
1	Agriculture	0.632220	0.191184	0.364908	0.632220	General
2	Coal selection	0.361569	0.302232	0.354112	0.361569	General
3	Oil and natural gas exploration	0.456020	0.152932	0.570985	0.570985	Leading
4	Metallic mine selection	0.472680	0.116770	0.301520	0.472680	General
5	Nonmetallic mine selection	0.376720	0.277887	0.424350	0.424350	Leading
6	Food and tobacco processing	0.523447	0.553857	0.363039	0.553857	Auxiliary
7	Textile	0.446160	0.488804	0.392865	0.488804	Auxiliary
8	Clothing, leather, plume and fibre products	0.357388	0.467901	0.551648	0.551648	Leading
9	Wood processing and furniture manufacturing	0.441209	0.698553	0.410898	0.698553	Auxiliary
10	Paper-making and printing, culture and educational goods	0.343203	0.819907	0.508344	0.819907	Auxiliary
11	Oil processing and coking	0.567046	0.536494	0.290253	0.567046	General
12	Chemical	0.367097	0.559472	0.463764	0.559472	Auxiliary
13	Nonmetallic mineral products	0.403300	0.447022	0.588130	0.588130	Leading
14	Metal smelting and rolling processing	0.131160	0.192840	0.866272	0.866272	Leading
15	Fabricated metal products	0.514348	0.520181	0.308697	0.520181	Auxiliary
16	Machinery	0.411720	0.439203	0.403664	0.439203	Auxiliary
17	Transportation equipment manufacturing	0.621571	0.321347	0.294834	0.621571	General
18	Electrical machinery and equipment manufacturing	0.406522	0.558490	0.572612	0.572612	Leading
19	Electronics and telecommunications equipment manufacturing	0.219140	0.234407	0.787729	0.787729	Leading
20	Instruments and office machinery manufacturing	0.300560	0.167885	0.716278	0.716278	Leading
21	other manufacturing	0.291620	0.337112	0.706755	0.706755	Leading
22	Electricity, steam and hot water production and supply	0.234600	0.507037	0.607478	0.607478	Leading
23	Gas production and supply	0.140000	0.085474	0.866738	0.866738	Leading
24	Water production and supply	0.802820	0.295565	0.183061	0.802820	General
25	Construction	0.300875	0.570662	0.640208	0.640662	Leading
26	Transportation, posts and telecommunications	0.538509	0.440473	0.269800	0.538509	General
27	Catering business	0.431846	0.368553	0.378152	0.431846	General
28	Finance and insurance	0.518053	0.406363	0.264314	0.518053	General
29	Real estate	0.435460	0.525231	0.465596	0.525231	Auxiliary

continued

Table 10.10 (continued) Integrated Clustering Evaluation Results of 34 Sectors in S Province

NO.	INDUSTRY	GENERAL INDUSTRY	AUXILIARY INDUSTRY	LEADING INDUSTRY	MAX	CLUSTERING RESULTS INDUSTRY
30	Social services	0.285857	0.817783	0.515315	0.817783	Auxiliary
31	Sanitation and physical education, social welfare sector	0.239831	0.379529	0.771005	0.771005	Leading
32	Education, arts, radio, film and television	0.437848	0.440584	0.319578	0.437848	Auxiliary
33	Comprehensive scientific research and technical service	0.397598	0.562166	0.417001	0.562166	Auxiliary
34	Executive authorities and other sectors	0.223860	0.369032	0.783755	0.783755	Leading

coking industry, transportation equipment manufacturing, water production and supply industry, transportation, posts and telecommunications industry, catering business, finance and insurance industry, the executive authorities and other sectors.

The leading industries are oil and natural gas exploration, oil processing and coking industry, chemical industry, nonmetallic mineral products industry, metal smelting and rolling processing industry, fabricated metal products, electronics and telecommunications equipment manufacturing, instruments and office machinery manufacturing, other manufacturing, electricity, steam and hot water production and supply, gas production and supply industry, construction industry, transportation, posts and telecommunications industry, real estate, social services, sanitation, physical education and social welfare sector, education, arts, radio, film and television industry, comprehensive scientific research and technical service, the executive authorities and other sectors. Which sectors are the priority selections among the leading industry candidates? We divide the whole industry into three categories (i.e, $s = 3$). The leading industries we calculated belong to the first grey type; therefore, the integrated decision-making measure formula is $\omega_i = 3\delta_i^1 + 2\delta_i^2 + \delta_i^1$. The integrated decision-making measure of each sector is shown in Table 10.11.

A comparison allows the leading industry candidates to be listed according to their integrated clustering measures as follows: other manufacturing, chemical, gas production and supply, social services, transportation, posts and telecommunications, fabricated metal products, sanitation, physical education and social welfare sector, nonmetallic

Table 10.11 Integrated Clustering Measure of Leading Industries in S Province

INDUSTRY	INTEGRATED CLUSTERING COEFFICIENT, δ_i^k			INTEGRATED CLUSTERING MEASURE, ω_i
	LEADING	AUXILIARY	GENERAL	
Oil and natural gas exploration	0.5249	0.2182	0.2569	2.268
Oil processing and coking	0.5412	0.3981	0.0607	2.4805
Chemical	0.7480	0.2266	0.0254	2.7226
Nonmetallic mineral products	0.5818	0.4182	0	2.5818
Fabricated metal products	0.6872	0.2714	0.0414	2.6458
Electronics and telecommunications equipment manufacturing	0.5020	0.4861	0.0119	2.4901
Instruments and office machinery manufacturing	0.6793	0.1987	0.1220	2.5573
Other manufacturing	0.7782	0.2218	0	2.7782
Electricity, steam and hot water production and supply	0.5997	0.3667	0.0336	2.5661
Gas production and supply	0.8591	0.0010	0.1399	2.7192
Construction	0.6858	0.1937	0.1205	2.5653
Transportation, posts and telecommunications	0.7081	0.2306	0.0613	2.6468
Real estate	0.5958	0.2739	0.1303	2.4655
Social services	0.6911	0.2806	0.0283	2.6628
Sanitation and physical education, social welfare sector	0.7163	0.1643	0.1194	2.5969
Education, arts, radio, film and television	0.7290	0.0800	0.1910	2.538
Comprehensive scientific research and technical service	0.6139	0.2444	0.1417	2.4722
The executive authorities and other sectors	0.6325	0.2454	0.1222	2.5105

mineral products, electricity, steam and hot water production and supply, construction, instruments and office machinery manufacturing, education, arts, radio, film and television, the executive authorities and other sectors, electronics and telecommunications equipment manufacturing, oil processing and coking, comprehensive scientific research and technical service, real estate, oil and natural gas exploration. S Province can refer to the above order when determining the leading industries.

One of the main tasks of S Province in economic development is to achieve the transformation of economic growth mode from extensive type to intensive type and the industrial structure upgrading. The key is the right choice and promotion of development of leading industries vigorously. The leading industries have a strong ability to absorb new high-tech technology and speed up the progress, as well as high correlation and penetration, which can reduce material and energy consumption, enhance the depth of processing and increase the added value, drive the upgrading of the whole industrial structure, and at the same time absorb many workers. The upgrading of industrial structure is a process of substitution of leading industries. The selection of leading industries is not a one-off choice, which has been developing and evolving continuously along with the regional economic development. At new stages of development, some of the old leading industries will be replaced by the new ones, and the regional economic structure also will be changed. As a result, when nurturing and developing the leading industries, S Province must make an overall plan and correctly handle the relationship between the development of industries and keeping thecoordinated development of regional economies.

10.3 S Direction and Keystone of the Industrial Structure Adjustment in S Province

10.3.1 Direction of the Industrial Structure Adjustment in the Three Major Industries in S Province

We make use of the dynamic input–output model of Leontief (Leontief "fast track" model) to adjust the industrial structure in three major industries in S Province and to make sure that the economic growth of S Province enters the "fast track" as quickly as possible to rationalize the industrial structure. We merge the 40×40 (or 40 sectors) input–output table in S Province in 1997 into the 3×3 (three industries) input–output table. The input–output data of the three major industries in S Province in 1997 is shown in Table 10.12.

The direct consumption coefficient matrix is obtained by the input–output tables:

$$A = \begin{bmatrix} 0.20462 & 0.02851 & 0.01194 \\ 0.25215 & 0.59639 & 0.27153 \\ 0.04272 & 0.13061 & 0.27775 \end{bmatrix}$$

Table 10.12 Input–Output Table of the ¯hree Major Industries in S Province in 1997 (100 million RMB)

		OUTPUT						
		INTERMEDIATE DEMAND						
INPUT		PRIMARY	SECONDARY	TERTIARY	FINAL CONSUMPTION	CAPITAL FORMATION	REGIONAL NET INFLOW	AGGREGATE DEMAND
Intermediate input	Primary industry	371.7	415.4	55.2	581.6	204.6	187.9	1816.4
	Secondary industry	458.0	8689.4	1254.6	1241.9	2520.2	405.9	14570.0
	Tertiary industry	77.6	1903.0	1283.4	1453.1	4.5	−101.1	4620.5
Depreciation and increment		909.1	3562.2	2027.4				
Total supply		1816.4	14570.0	4620.5				

In calculating the corresponding fixed assets usage coefficient matrix, both b_{1j} and b_{3j} ($j = 1, 2, 3$) are equal to zero as the primary industry and tertiary industry do not form fixed assets. The total value of fixed assets of various sectors cannot be obtained directly from the statistical yearbook; therefore, the matrix can only be calculated through an indirect method. The fixed assets value of construction in S Province in 1997 is 24,280 million RMB, the fixed assets value per 100 independent accounting industries in the whole society is 175.92 RMB. The total output value of the secondary industry is 1457 billion RMB, and the output value of construction industry in S Province in the same period is 153,070 million RMB; then, the fixed assets value of the industry in S Province in 1997 is obtained through calculation as follows: $(14{,}570 - 1530.7)/1.7592 = 7412$. Thus, the fixed assets value of the secondary industry in S Province is $(7412 + 242.8)$ million RMB = 76tn RMB.

On the basis of the total fixed assets value of the secondary industry, the respective total value of fixed assets in the primary industry and tertiary industry can be calculated according to the ratio of fixed assets value of the primary, secondary, and tertiary industries. The ratio of fixed assets value in primary, secondary, and tertiary industry can be replaced by the formation ratio of fixed assets of the three major industries, so the ratio of fixed assets investment of the primary, secondary, and tertiary industries in S Province can be replaced by the average value of the formation ratio of fixed assets from 1992 to 1997 which are 2.452%, 48.472%, and 49.076% respectively. Therefore, the total value of fixed assets of the primary industry is 38,720 million RMB and that of the tertiary industry is 7750.2 RMB. The fixed assets usage coefficient matrix B is given by

$$B = \begin{bmatrix} 0 & 0 & 0 \\ 0.21317 & 0.52538 & 1.67735 \\ 0 & 0 & 0 \end{bmatrix}$$

Then, we next calculate the consumption coefficient matrix. The respective number of labor force of the primary, secondary, and tertiary industries in S Province in 1997 is 15,542,200; 12,191,500; and

8,846,800. The consumption coefficient matrix can be obtained as follows:

$$T = \begin{bmatrix} 0.13618 & 0.01339 & 0.03056 \\ 0.29269 & 0.03048 & 0.06615 \\ 0.34711 & 0.040632 & 0.08321 \end{bmatrix}$$

Suppose the economic growth rateα of each sector is the same, namely,

$$x(t+1) = (1+\alpha)x(t) \tag{10.1}$$

Put Equation (1) into the Leontief "fast track" model we get,

$$x(t) = Ax(t) + B[x(t+1) - x(t)] + Tx(t)$$

Then,

$$\begin{aligned} x(t) &= Ax(t) + B[(1+\alpha)x(t) - x(t)] + Tx(t) \\ &= Ax(t) + \alpha Bx(t) + Tx(t) \end{aligned}$$

So,

$$\frac{1}{\alpha}x(t) = (I - A - T)^{-1}Bx(t) \tag{10.2}$$

and $H = (I - A - T)^{-1}\ B$. Then, Equation 10.2 can be rewritten as follows: $(1/\alpha)x(t) = Hx(t)$.

We find from the above formulas that the balanced growth rate of Leontief dynamic input–output model α is the reciprocal of the eigenvalue $1/\alpha$ of matrix H, and that the output vector $x(t)$ is the corresponding unit eigenvalue vector.

First of all, we calculate matrix H.

$$H = (I - A - T)^{-1}B = \begin{bmatrix} 0.0934 & 0.2229 & 0.7035 \\ 1.1103 & 2.6494 & 8.3605 \\ 0.3345 & 0.8460 & 2.6695 \end{bmatrix}$$

The economic growth rate and output structure of the three major industries in S Province can be obtained through calculation. The eigenvalue $1/\alpha$ can be derived from matrix H, which is 5.538. Then the growth rate of economic balance a equals 18% ($a = 1/5.538 = 0.1801 = 18\%$). The unit eigenvalue vector of matrix H is the industrial structure of the primary, secondary, and tertiary industries, respectively, which are 5.99%, 70.57%, and 23.44%.

In fact, the industrial structure of the total social production value of S Province in 1997 is (9.17%, 73.24%, 17.59%), based on the statistical yearbook. This shows that there is a certain gap between the industrial structure of S Province in 1997 and the balanced growth path. According to the results, the growth rate of balance is 18%, which appears to be on the high side, but compared with the development speed in the past, we think the speed is suitable for S Province; because the average growth rate of the total social production value of S Province from 1992 to 2000 is 22.2%, the economic growth of S Province has declined due to the Asian financial crisis in 1997 and the domestic industrial structure adjustment, but S Province can reach an annual growth rate of economic balance of 18% through its effort.

As the track of balanced growth is the best track, that is to say, if the output proportion of economic system is just on the balanced growth path, it should go ahead along the path. However, if the proportion of economic output is unreasonable in the initial period, which means the industrial structure is not on the balanced path, the industrial structure should be adjusted to the balanced growth path as soon as possible, and develop along the path. Then we carry out the regional economic industrial structure adjustment of S Province by the linear programming model.

According to previous discussions, we see that the reasonable industrial structure in 1997 should be 5.99%, 70.57%, and 23.44%, respectively; in fact, the industrial structure was 9.17%, 73.24% and 17.59%, that is to say, the industrial structure is not on the track of balanced growth. Thereby, an appropriate adjustment should be carried out as soon as possible to adjust the industrial structure to the "fast track." If the immediate adjustment is not realistic, we plan to use eight years doing this, hoping that the economic structure will be adjusted to a reasonable track. In other words, the industrial structure of S Province after eight years should be as shown below (structure calculated in accordance with the output value):

$$x(8) = \begin{bmatrix} 5.99\% \times q \\ 70.57\% \times q \\ 23.44\% \times q \end{bmatrix}$$

where q is an unknown parameter, that is to say, after eight years, the ratio of the industrial structure of the three major industries in S Province is 5.99%, 70.57%, and 23.44%, respectively, and if we want to make the output as large as possible, we should make q large enough.

As the total demand each year is smaller than the total available output, there are obviously constraints such as: $Ax(t) + B[x(t+1) - x(t)] + Tx(t) \leq x(t)$ where $Ax(t)$ is the intermediate input demand, $B[x(t+1) - x(t)]$ is the investment demand, $Tx(t)$ is the final consumer demand and $x(t)$ in the output.

Based on the above analysis, the linear programming model is as follows:

$$\max q$$

$$\begin{cases} Ax(0) + B[x(1) - x(0)] + Tx(0) \leq x(0) \\ Ax(1) + B[x(2) - x(1)] + Tx(1) \leq x(1) \\ \vdots \\ Ax(7) + B[x(8) - x(7)] + Tx(7) \leq x(7) \\ \bar{x} \cdot q \leq x(8), \quad q > 0 \\ x(1), x(2), \ldots, x(8) \geq 0 \end{cases}$$

where A, B, and T are as mentioned above, respectively. $x(0)$ the industrial structure in the initial period, that is, in 1997, is given by $x(0) = (9.17\%, 73.24\%, 17.59\%)^T$ and $\bar{x}$ is the optimal industrial structure, which is the industrial structure on the "fast track": $\bar{x} = (5.99\%, 70.57\%, 23.44\%)^T$. The industrial structure $x(1), x(2), \ldots, x(8)$ are of

1998, 1999, …, 2005, respectively, and all $x(1)$, $x(2)$, …, $x(8)$ are three-dimensional column vectors.

We solve the above linear programming model by using software. The result is to enter the "fast track" in the first year, which is clearly inconsistent with the actual situation. Then we consider increasing some constraints. The balanced growth rate of each industry is 18% which is believed to be inappropriate for the primary industry; the primary industry being the foundation industry of the national economy is believed to have an output growth rate of not less than 8%; the speed of development of the tertiary industry should be greater than that of the whole national economy, but the output growth rate should not be larger than 25%. After adding the above constraints, we solve the linear programming model through computer simulation, and then normalize the results. The corresponding industrial structure of the total social production value of each industry is shown in Table 10.13. From the table it can be seen that after the eight years adjustment, S Province has adjusted its industrial structure to the best state, and has made its output reach the maximum value in the mentioned structure. And the above result is the optimal output structure of the total social output value.

The above-mentioned structure is the industrial structure of the total social output value; however, what we normally need is the industrial structure of GDP. Therefore, we will transform the mentioned structure into the industrial structure of GDP.

We make the average value of the ratio of the GDP of each industry and the total social output value from 1994 to 2000 as the conversion factors. The conversion factors of the primary, secondary, and tertiary

Table 10.13 Prediction of the Total Social Production Value Structure of Each Industry in S Province (%)

YEAR	PRIMARY	SECONDARY	TERTIARY
1998	8.75	72.88	18.37
1999	8.32	72.52	19.16
2000	7.81	72.14	20.05
2001	7.43	71.77	20.80
2002	7.05	71.46	21.49
2003	6.64	71.14	22.22
2004	6.26	70.82	22.92
2005	5.99	70.57	23.44

Table 10.14 Comparison between Analog Value and Actual Value in the Three Major Industries in S Province (%)

	PRIMARY INDUSTRY		SECONDARY INDUSTRY		TERTIARY INDUSTRY	
YEAR	ANALOG VALUE	ACTUAL VALUE	ANALOG VALUE	ACTUAL VALUE	ANALOG VALUE	ACTUAL VALUE
1998	14.0	14.1	51.5	50.6	34.5	35.3
1999	13.6	13.0	51.0	50.9	35.4	36.1
2000	12.7	12.0	50.4	51.6	36.9	36.4

industries are 0.4974, 0.2132, and 0.5603, respectively. The GDP structure adjustment program is shown in Tables 10.14 and 10.15.

From Table 10.14 we find that the industrial structure available by the "fast track" model is basically the same as the actual industrial structure, which shows that it is feasible to adopt the "fast track" model to adjust the industrial structure. The adjustment program of S Province is shown in Table 10.15. After the economic structure adjustment of five years, the respective proportion that the primary and secondary industry accounted for is gradually declining, while the proportion of the tertiary industry is rising step by step. The industrial structure of the primary industry decreased from 12% in 2000 to 9.6% in 2005, decreasing by 2.4%; that of the secondary industry drops from 51.6% in 2000 to 48.3% in 2005, with a decrease of 3.3%; and at the same time, that of the tertiary industry rose from 36.4% in 2000 to 42.1% in 2005, with an increase of 5.7%. After efforts of several years, the economic structure of S Province will change from the current "two, three, one" structure to the "three, two, one" structure.

The industrial structure calculated through the "fast track" model is being adjusted continuously with the technological progress, that is to say, it changes with the consumption coefficient matrix A, the fixed

Table 10.15 Internal Industrial Restructuring Program in the Three Major Industries in S Province (%)

YEAR	PRIMARY INDUSTRY	SECONDARY INDUSTRY	TERTIARY INDUSTRY
2001	12.05	49.92	37.99
2002	11.39	49.49	39.12
2003	10.68	49.05	40.27
2004	10.03	48.62	41.35
2005	9.56	48.29	42.15

assets usage coefficient matrix B and consumption coefficient matrix T in the input–output table. As a result, with the economic development, we should be ready to adjust the balanced growth path so that the development of the economic structure always runs along the "fast track."

10.3.2 Direction of the Internal Industrial Structure Adjustment in the Primary Industry in S Province

In the face of domestic relative surplus of agricultural products, increased market constraints and slowing growth in the income of farmers, the irrational industrial structure of agriculture has become one of the major factors affecting the rural economic development and restricting the income of farmers; therefore, the internal structure adjustment in agriculture is an effective way to solve the problem of slowing growth in the income of farmers.

Through the internal structure analysis of the primary industry, the farm industry is divided into three types: food crops, other cash crops, and vegetables and fruits; the animal husbandry industry is also divided into three types: domestic animals, poultry, and other animals. In this way, the primary industry is divided into the nine sectors: food crops, other cash crops, vegetables and fruits, other agriculture, forestry, domestic animals, poultry, other animals, and fisheries. After analysis, we believe that there is a linear relationship among the output value of food crops, other cash crops, vegetables and fruits, and other agriculture. Then to make it a linear regression equation:

$$\mathrm{LS} = 2.7176\mathrm{QZ} - 1.799\mathrm{SC} + 1.4099\mathrm{QN} + 136.52$$

where LS is the output value of food crops, QZ is the output value of other cash crops, SC is the output value of vegetables and fruits, and QN is the output value of other agriculture.

Similarly, there is also a linear relationship between the output value of other cash crops and the value of vegetables and fruits. This we get through linear regression:

$$\mathrm{SC} = 1.2405\mathrm{QZ} - 11.19$$

The relationship between the output value of food crops and three indicators of the animal husbandry is

$$\mathrm{LS} = 1.8965\mathrm{XL} + 0.7941\mathrm{QL} - 5.8584\mathrm{QD} + 175.5$$

where XL is the output value of domestic animals, QL is the output value of poultry, and QD is the output value of other animals.

The relationship among the output value of fisheries and three indicators in the animal husbandry is given by

$$\mathrm{YE} = 0.6161\mathrm{XL} + 0.6682\mathrm{QL} + 1.7733\mathrm{QD} - 60.66$$

where YE is the output value of fisheries.

The relationship between domestic animal and poultry is given by

$$\mathrm{XL} = 0.5963\mathrm{QL} + 44.82$$

Through the above analysis, the linear programming model is as follows:

$$\max Q_{2001} = \mathrm{LS} + \mathrm{QZ} + \mathrm{SC} + \mathrm{QN} + \mathrm{LY} + \mathrm{XL} + \mathrm{QL} + \mathrm{QD} + \mathrm{YE}$$

$$\text{s.t.} \begin{cases} \mathrm{LS} - 2.7176\mathrm{QZ} + 1.799\mathrm{SC} - 1.4099\mathrm{QN} = 136.52 \\ -\mathrm{SC} + 1.2405\mathrm{QZ} = 11.19 \\ \mathrm{LS} - 1.8965\mathrm{XL} - 0.7941\mathrm{QL} + 5.8584\mathrm{QD} = 175.5 \\ 0.6161\mathrm{XL} + 0.6682\mathrm{QL} + 1.7733\mathrm{QD} - \mathrm{YE} = 60.66 \\ \mathrm{QL} - 0.5963\mathrm{QL} = 44.82 \\ \mathrm{LS} + \mathrm{QZ} + \mathrm{SC} \le 901.95 \\ \mathrm{XL} + \mathrm{QL} + \mathrm{QD} \le 477.3 \\ 392.7 \le \mathrm{LS} \le 402.4 \\ 183.0 \le \mathrm{QZ} \le 190.1 \\ 317.5 \le \mathrm{SC} \le 338.6 \\ 251.1 \le \mathrm{QN} \le 260.6 \\ 32.3 \le \mathrm{LY} \le 33.2 \\ 211.1 \le \mathrm{XL} \le 217.4 \\ 193.3 \le \mathrm{QL} \le 204.2 \\ 56.1 \le \mathrm{QD} \le 59.6 \\ 341.2 \le \mathrm{YE} \le 359.9 \end{cases}$$

where LY is the output value of forestry, Q is the total output value of the primary industry.

Through computer calculations, the results are

$$LS = 392.7,\ QZ = 183.0,\ SC = 326.2,\ QN = 260.6,\ LY = 33.2$$

$$XL = 213.5,\ QL = 204.2,\ QD = 59.6,\ YE = 359.9$$

Then we do appropriate adjustments in the latter 11 constraints of the above linear programming, and ensure that food production remains stable; the results are as follows:

$$Q^*_{2002} = 2206.1$$

$$LS = 402.2,\ QZ = 192.9,\ SC = 361.0,\ QN = 281.4,\ LY = 36.1$$

$$XL = 231.6,\ QL = 230.4,\ QD = 67.5,\ YE = 403.1$$

$$Q^*_{2003} = 2366.9$$

$$LS = 410.2,\ QZ = 201.6,\ SC = 391.9,\ QN = 281.4,\ LY = 38.6,$$

$$XL = 252.4,\ QL = 258.3,\ QD = 77.0,\ YE = 455.4$$

$$Q^*_{2004} = 2553.7$$

$$LS = 420.6,\ QZ = 209.0,\ SC = 424.2,\ QN = 301.9,\ LY = 41.6,$$

$$XL = 277.2,\ QL = 2082.3,\ QD = 86.1,\ YE = 510.7$$

$$Q^*_{2005} = 2747.2$$

$$LS = 431.0,\ QZ = 218.4,\ SC = 461.5,\ QN = 326.0,\ LY = 44.9,$$

$$XL = 302.1,\ QL = 311.1,\ QD = 96.3,\ YE = 555.9$$

From Table 10.16 it can been seen that through five years of adjustment in agriculture, the ratio of agriculture, forestry, animal

Table 10.16 Prediction in the Internal Restructuring in the Primary Industry in S Province (%)

	2000	2001	2002	2003	2004	2005
Ratio of agriculture, forestry, animal husbandry, and fisheries	58.6:1.6: 23.0:16.8	57.2:1.63: 23.5:17.7	56.1:1.6: 24:18.3	54.5:1.6: 24.7:19.2	53.1:1.6: 25.3:20	52.3:1.6: 25.8:20.3
Proportion of faming in the primary industry	46	44.3	43.3	42.4	41.3	40.4
Ratio of food crop and cash crops	45.3:54.7	43.5:56.5	42.1:57.9	41:59	39.9:60.1	38.8:61.2
Output value (100 million RMB)	1869.7	2033	2206.1	2366.9	2553.7	2747.2
Growth speed of output value		8.77	8.5	7.95	7.9	7.9

husbandry, and fisheries changed from 58.6%:1.6%:23.0%:16.8% in 2000 to 52.3%:1.6%:25.8%:20.3% in 2005. The proportion of agriculture in the primary industry dropped by 6.3%, while the proportion of animal husbandry and fisheries increased by 2.8% and 3.5%, respectively. At the same time, the proportion of farming in the primary industry dropped from 46% in 2000 to 40.4% in 2005, while the ratio of food crops and cash crops also changed from 45.3%:54.7% in 2000 to 38.8%:61.2% in 2005. All these indicate that the internal structure of the primary industry gradually becomes rational.

10.3.3 Direction of the Internal Industrial Structure Adjustment in the Secondary Industry in S Province

We calculate the adjustment direction in 1997 using the input–output table. First of all, the secondary industry is divided into eight sectors, namely mining, food, textile, clothing and paper-making, petrochemical, machinery building and equipment, urban public utilities, construction. The primary industry and tertiary industry are looked as two independent sectors. Then we calculate through Leontief "fast track" model.

In the calculation, we merge the 40 × 40 input–output table into the 10 × 10 input–output table (shown in Table 10.17).

Table 10.17 The 10 × 10 Input–Output Table of S Province in 1997 (100 million RMB)

INPUT \ OUTPUT		INTERMEDIATE DEMAND													
		PRIMARY	MINING	FOOD	TEXTILE	CLOTHING AND PAPER-MAKING	PETROLEUM AND CHEMICAL	MACHINERY BUILDING AND EQUIPMENT	URBAN PUBLIC UTILITIES	CON-STRUCTION	TERTIARY	ULTIMATE CONSUMPTION	CAPITAL FORMATION	REGIONAL NET INFLOW	AGGREGATE DEMAND
Intermediate input	Primary	371.7	0.2	321.4	62.3	16.7	13.1	1.7	0	0	55.2	58.2	204.6	187.9	1816.4
	Mining	2.6	25.0	5.1	7.2	52.7	155.5	178.8	119.6	109.4	33.1	0.2	15.7	–477.3	227.8
	Food	136.4	0.0	353.0	0.1	17.4	19.2	0.2	0	0	113.7	525.3	52.0	1.6	1219.0
	Textile	7.9	0.3	1.5	723.0	194.6	25.8	23.7	0.2	0	7.6	62.4	78.2	331.7	1456.9
	Clothing and paper-making	7.2	1.6	22.9	5.6	297.6	22.6	78.9	2.7	37.9	144.2	193.5	33.9	446.0	1294.6
	Petroleum and chemical	176.4	14.1	31.9	138.0	85.5	966.5	320.0	16.6	43.4	284.2	80.1	75.0	45.2	2276.9
	Machinery building and equipment	71.9	30.2	38.8	32.3	77.2	99.4	3089.2	50.9	698.2	424.2	304.8	968.1	–3.1	5882.2
	Urban public utilities	48.0	13.0	15.8	35.8	35.1	78.9	215.1	54.4	17.1	121.9	75.7	1.2	–29.9	682.0
	Construction	7.7	0.5	0.5	0.5	0.8	1.1	4.2	1.4	0.5	125.7	0	1296.1	91.8	1530.7
	Tertiary	77.6	30.0	141.5	143.6	173.1	323.4	775.4	123.6	192.3	1283.4	1453.1	4.5	–101.1	4620.5
	Total intermediate investment	907.3	114.8	932.6	1148.4	950.9	1705.5	4687.3	369.4	1098.8	2593.1				
Depreciation and increment		909.1	113.0	286.4	308.5	343.7	571.3	1194.9	312.6	431.9	2027.4				
Total supply		1816.4	227.8	1219.0	1456.9	1294.6	2276.8	5882.2	682.0	1530.7	4620.5				

And we calculate the direct consumption coefficient matrix A by the input–output table

$$A = \begin{bmatrix} 0.2046 & 0.0008 & 0.2637 & 0.0427 & 0.0129 & 0.0058 & 0.0003 & 0 & 0 & 0.0119 \\ 0.0014 & 0.1096 & 0.0042 & 0.0050 & 0.0407 & 0.0683 & 0.0304 & 0.1753 & 0.0715 & 0.0072 \\ 0.0752 & 0 & 0.2896 & 0.0001 & 0.0135 & 0.0084 & 0 & 0 & 0 & 0.0246 \\ 0.0436 & 0.0012 & 0.0012 & 0.4963 & 0.1503 & 0.0113 & 0.0040 & 0.0003 & 0 & 0.0016 \\ 0.0040 & 0.0068 & 0.0188 & 0.0038 & 0.2299 & 0.0099 & 0.0134 & 0.0040 & 0.0248 & 0.0312 \\ 0.0971 & 0.0620 & 0.0262 & 0.0947 & 0.0660 & 0.4245 & 0.0544 & 0.0243 & 0.0284 & 0.0615 \\ 0.0396 & 0.1327 & 0.0319 & 0.0222 & 0.0597 & 0.0436 & 0.5252 & 0.0746 & 0.4561 & 0.0918 \\ 0.0264 & 0.0571 & 0.0130 & 0.0246 & 0.0271 & 0.0347 & 0.0366 & 0.0798 & 0.0112 & 0.0264 \\ 0.0042 & 0.0021 & 0.0004 & 0.0003 & 0.0006 & 0.0047 & 0.0007 & 0.0020 & 0.0003 & 0.0272 \\ 0.0427 & 0.1318 & 0.1161 & 0.0986 & 0.1337 & 0.1420 & 0.1318 & 0.1813 & 0.1257 & 0.2778 \end{bmatrix}$$

And then we determine the fixed assets usage coefficient matrix B and consumption coefficient matrix T. By the previous calculation, we know that the value of industrial fixed assets is 741,200 million RMB; the value of fixed assets in construction industry is 24,280 million RMB.

In capital formation of the secondary industry, only machinery building and equipment industry and construction industry can form fixed assets. Therefore, we have to divide the fixed assets of each sector into two parts: the formation fixed assets of machinery building and equipment industry and the formation fixed assets of construction industry. First of all, we calculate the fixed assets of each sector, and use the ratio of original value of fixed assets in the sub-sector of independent accounting enterprises in place of fixed assets of sub-sectors, and then we divide the total fixed assets investment into construction and installation project and the purchase of equipment. We know from the statistical yearbook that the ratio of the two sectors is 0.655:0.345. We take construction and installation project as fixed assets provided by construction industry, the purchase of equipment as fixed assets provided by machinery building and equipment industry. Therefore, the fixed assets of each sector in industry can be divided into two types; one is provided by construction industry and the other by machinery building and equipment. For fixed assets of construction industry, it is divided into the fixed assets for self-owned equipment and the fixed assets for construction usage, which are also provided by machinery building and equipment and construction industry. The fixed assets of the primary industry and that of the tertiary industry are still divided into construction and installation project and the purchase of equipment in accordance with the ratio of 0.655:0.345. The detailed data are shown in Table 10.18.

Table 10.18 Constitution of Fixed Assets of Each Sector (100 million RMB)

INDUSTRY	CONSTRUCTION AND INSTALLATION WORKS	PURCHASE OF EQUIPMENT
Primary	253.6	133.6
Mining	150.2	78.1
Food	269.8	142.1
Textile	604.6	318.4
Clothing and paper-making	296.1	155.9
Petrochemical	1333.6	702.4
Machinery building and equipment	1568.1	825.9
Urban public utilities	630.2	331.9
Construction	150.0	92.8
Tertiary	5076.4	2673.8

Therefore, the fixed assets usage coefficient matrix B is as follows:

$$B = \begin{bmatrix} 0 & 0 & 0 & 0 & 0 & 0 & 0 & 0 & 0 & 0 \\ 0 & 0 & 0 & 0 & 0 & 0 & 0 & 0 & 0 & 0 \\ 0 & 0 & 0 & 0 & 0 & 0 & 0 & 0 & 0 & 0 \\ 0 & 0 & 0 & 0 & 0 & 0 & 0 & 0 & 0 & 0 \\ 0 & 0 & 0 & 0 & 0 & 0 & 0 & 0 & 0 & 0 \\ 0 & 0 & 0 & 0 & 0 & 0 & 0 & 0 & 0 & 0 \\ 0.0736 & 0.3431 & 0.1166 & 0.2186 & 0.2046 & 0.3086 & 0.1404 & 0.4866 & 0.0606 & 0.5787 \\ 0 & 0 & 0 & 0 & 0 & 0 & 0 & 0 & 0 & 0 \\ 0.1397 & 0.6596 & 0.2214 & 0.4150 & 0.2084 & 0.5857 & 0.2666 & 0.9240 & 0.0980 & 1.0988 \\ 0 & 0 & 0 & 0 & 0 & 0 & 0 & 0 & 0 & 0 \end{bmatrix}$$

The ratio of labor force in the primary, secondary, and tertiary industries is 42.5%, 33.3%, and 24.2%, respectively, while in the secondary industry, the labor force of construction industry accounts for 7.72% in the total labor force of the society, the labor force proportions of other sectors in the total labor force of the society can be calculated according to output value and labor productivity of the independent accounting industries as follows: mining industry 0.72%, food industry 1.43%, textile industry 4.26%, clothing and paper-making industry 2.81%, petrochemical industry 6.19%, machinery building and equipment 9.65%, urban public utilities 0.53%, and construction industry 7.72%. Then the consumption coefficient matrix T is as follows:

$$T = \left[\begin{matrix} 0.1361 & 0.0184 & 0.0068 & 0.0057 & 0.1262 \\ 0.0001 & 0 & 0 & 0 & 0 \\ 0.1229 & 0.0166 & 0.0062 & 0.0154 & 0.0114 \\ 0.0146 & 0.0020 & 0.0007 & 0.0018 & 0.0214 \\ 0.0453 & 0.0061 & 0.0023 & 0.0057 & 0.0042 \\ 0.0187 & 0.0233 & 0.0009 & 0.0023 & 0.0017 \\ 0.0713 & 0.0058 & 0.0036 & 0.0089 & 0.0066 \\ 0.0177 & 0.0014 & 0.0009 & 0.0022 & 0.0016 \\ 0 & 0 & 0 & 0 & 0 \\ 0.3400 & 0.0276 & 0.0171 & 0.0425 & 0.0315 \end{matrix}\right.$$

$$\left.\begin{matrix} 0.0158 & 0.0095 & 0.0045 & 0.0293 & 0.0355 \\ 0 & 0 & 0 & 0 & 0 \\ 0.0143 & 0.0086 & 0.0041 & 0.0265 & 0.0275 \\ 0.0017 & 0.0010 & 0.0005 & 0.0031 & 0.0033 \\ 0.0053 & 0.0032 & 0.0015 & 0.0098 & 0.0101 \\ 0.0022 & 0.0013 & 0.0006 & 0.0040 & 0.0042 \\ 0.0083 & 0.0050 & 0.0024 & 0.0154 & 0.0160 \\ 0.0021 & 0.0012 & 0.0006 & 0.0038 & 0.0040 \\ 0 & 0 & 0 & 0 & 0 \\ 0.0395 & 0.0238 & 0.0113 & 0.0733 & 0.0761 \end{matrix}\right]$$

By matrix A, B, T, and Leontief "fast track" model,

$$\frac{1}{\alpha}x(t) = Hx(t)$$

where $H = (I - A - T)^{-1} B$. Then we can derive $\alpha = 18.9\%$.

The unit eigenvalue vector of matrix H is as follows:

(5.75%, 3.24%, 4.74, 2.6%, 3.29%, 9.45%, 32.7%, 3.44%, 11.8%, 22.97%).

Table 10.19 Actual Structure and Balanced Structure of Each Sector in S Province in 1997 (%)

INDUSTRY	ACTUAL STRUCTURE	BALANCED STRUCTURE
Primary	9.17	5.75
Mining	1.14	3.24
Food	6.12	4.74
Textile	7.33	2.6
Clothing and paper-making	6.5	3.29
Petrochemical	11.45	9.45
Machinery building and equipment	29.57	32.7
Urban public utilities	3.43	3.44
Construction	7.7	11.8
Tertiary	17.59	22.97

This is the optimal industrial structure in the state of the balanced growth rate. Put the unit eigenvalue vector of matrix H into $(1/\alpha)x(t) = Hx(t)$.

We know from the statistical yearbook in S Province in 1997 that the actual structure of each sector is as given below (Table 10.19):

(9.17%, 1.14%, 6.12%, 7.33%, 6.50%, 11.45%, 29.57%, 3.43%, 7.70%, 17.59%)

By means of linear dynamic programming model, the industrial structure of all sectors of the primary, secondary, and tertiary industries can be calculated in accordance with the total social output value (shown in Table 10.20). Taking actual output value of industrial structure of each sector in 1997 as the initial condition of the linear dynamic programming model, the eigenvalue vector of the matrix H, namely, taking the optimal industrial structure as the final constraints, we can obtain the industrial structure of each sector from 1998 to 2005.

The ratio of the GDP of each sector and total social output value is considered as the conversion coefficient; then, the conversion coefficients of the primary industry, mining industry, food industry, textile industry, clothing and paper-making industry, petrochemical industry, machinery building and equipment, urban public utilities, construction industry and the tertiary industry are 0.4974, 0.4956, 0.2346, 0.2115, 0.2650, 0.2508, 0.2029, 0.4575, 0.2822 and 0.5603, respectively. The prediction of the GDP structure of each sector, namely the economic balanced growth path is shown in Table 10.21.

Table 10.20 Industrial Structure of Each Sector in S Province Calculated in Accordance with the Total Social Output Value (%)

INDUSTRY	1998	1999	2000	2001	2002	2003	2004	2005
Primary	8.72	8.25	7.71	7.28	6.88	6.44	6.02	5.75
Mining	1.41	1.69	1.96	2.23	2.56	2.78	3.03	3.24
Food	5.94	5.75	5.58	5.4	5.22	5.05	4.89	4.74
Textile	6.71	6.1	5.49	4.89	4.3	3.71	3.13	2.6
Clothing and paper-making	6.06	5.63	5.21	4.8	4.39	3.99	3.63	3.29
Petrochemical	11.19	10.92	10.66	10.41	10.17	9.93	9.69	9.45
Machinery building and equipment	29.99	30.41	30.78	31.17	31.52	31.93	32.32	32.7
Urban public utilities	3.43	3.43	3.43	3.43	3.43	3.43	3.43	3.44
Construction	8.23	8.76	9.28	9.78	10.29	10.82	11.34	11.8
Tertiary	18.32	19.06	19.9	20.61	21.24	21.92	22.52	22.97

10.3.4 Direction of Internal Restructuring in the Tertiary Industry in S Province

We calculate the adjustment direction by the use of the input–output table of 1997. First of all, the tertiary industry is divided into four sectors, namely the circulation sectors (including transportation, posts and telecommunications industry, commerce, material supply and marketing and warehousing industry), sectors for production and living services (including finance, insurance, geological survey industry, real estate, public utilities, services for residents, inquiry services,

Table 10.21 Industrial Structure of Each Sector Calculated in Accordance with GDP (%)

INDUSTRY	1998	1999	2000	2001	2002	2003	2004	2005
Primary	14.4	13.5	12.5	11.8	11.1	10.3	9.6	9.1
Mining	2.0	2.4	2.8	3.1	3.6	3.9	4.2	4.4
Food	4.1	3.9	3.8	3.6	3.5	3.3	3.2	3.1
Textile	4.1	3.7	3.3	3	2.6	2.2	1.9	1.5
Clothing and paper-making	4.7	4.3	4.0	3.6	3.3	3.0	2.7	2.4
Petrochemical	8.0	7.8	7.5	7.3	7.1	6.9	6.7	6.5
Machinery building and equipment	17.7	17.8	17.9	18	18.1	18.3	18.4	18.5
Urban pubic utilities	4.4	4.4	4.3	4.3	4.3	4.3	4.3	4.2
Construction	6.7	7.1	7.4	7.8	8.1	8.5	8.9	9.2
Tertiary	33.9	35.1	36.4	37.5	38.4	39.4	40.3	40.9

the integrated technology services, agriculture, forestry, animal husbandry, fisheries, water and water services sector, roads, inland waterways conservation, etc.), sectors for enhancing the scientific and cultural level and residents quality (including education, culture, radio and television, science research, sanitation, sports, and social welfare, etc.), and sectors for social public needs (including the state organs, party and government organs, social groups, the armed forces, and the police, etc.). The primary industry and the secondary industry are considered as two independent sectors; we then calculate using Leontief "fast track" model.

In the calculation, we merge the 40×40 input–output table into the 6×6 input–output table, shown in Table 10.22.

The direct consumption coefficient matrix A is obtained by the input–output table.

$$A = \begin{bmatrix} 0.2046 & 0.0285 & 0.0177 & 0.0044 & 0.0022 & 0.0260 \\ 0.2522 & 0.5970 & 0.1964 & 0.2593 & 0.4731 & 0.3671 \\ 0.0138 & 0.0946 & 0.1746 & 0.0639 & 0.0311 & 0.0817 \\ 0.0056 & 0.0025 & 0.1579 & 0.1502 & 0.0296 & 0.0503 \\ 0.0035 & 0.0045 & 0.0065 & 0.0114 & 0.0748 & 0.0311 \\ 0.0198 & 0.0012 & 0.0091 & 0.0297 & 0.0141 & 0.0413 \end{bmatrix}$$

We determine the fixed assets usage coefficient matrix B and consumption coefficient matrix T. We know from the previous calculation that the fixed assets value of the tertiary industry is 775,020 million RMB. Then we average the five years' value of infrastructure and investment for upgrading the fixed assets of the independent accounting enterprises, and the investment proportion of the four levels in the tertiary industry can be calculated by the average value: the first level is 55.36%, the second level 16.28%, the third level 12.02%, and the fourth level 16.34%. The fixed assets of the four levels occupied are 42,905,000 RMB, 12,617,000 RMB, 9,316,000 RMB, and 12,664,000 RMB, respectively; the fixed assets of the primary industry and secondary industry are 3,872,000 RMB and 76,548,000 RMB, respectively. As the primary industry and tertiary industry do

Table 10.22 Input–Output Table of the Three Major Industries in S Province in 1997 (100 million RMB)

INPUT \ OUTPUT		INTERMEDIATE DEMAND										
		PRIMARY	SECONDARY	FIRST LEVEL	SECOND LEVEL	THIRD LEVEL	FOURTH LEVEL	TOTAL INTERMEDIATE DEMAND	ULTIMATE CONSUMPTION	CAPITAL FORMATION	INTER-REGIONAL INFLOW AND OUTFLOW	AGGREGATE DEMAND
Intermediate input	Primary	371.7	415.4	36.3	6.2	1.5	11.2	842.2	581.6	204.6	187.9	1816.4
	Secondary	458.0	8689.4	416.2	363.1	316.9	158.3	10402.0	1241.9	2520.2	405.9	14570.0
	First level	25.2	1451.0	370.0	89.4	20.8	35.2	1991.7	221.5	4.5	−98.7	2118.9
	Second level	10.2	368.2	334.6	210.4	19.8	21.7	965.0	440.6	0	−5.3	1400.3
	Third level	6.3	65.9	13.7	16.0	50.1	13.4	165.4	491.9	0	13.9	669.9
	Fourth level	35.9	17.9	19.2	41.6	9.4	17.8	141.9	299.0	0	−10.9	431.4
	Total intermediate input	907.3	11007.8	1190.1	726.7	418.6	257.7	14508.2	3276.6	2729.3	492.7	21006.9
Depreciation and increment		909.1	3562.2	928.9	673.6	299.1	125.8	6498.7				
Total supply		1816.4	14570.0	2118.9	1400.3	669.9	431.4	21006.9				

not form fixed assets, the fixed assets usage coefficient matrix B is as follows:

$$B = \begin{bmatrix} 0 & 0 & 0 & 0 & 0 & 0 \\ 0.4609 & 0.5254 & 2.0248 & 0.9012 & 1.3907 & 2.9355 \\ 0 & 0 & 0 & 0 & 0 & 0 \\ 0 & 0 & 0 & 0 & 0 & 0 \\ 0 & 0 & 0 & 0 & 0 & 0 \\ 0 & 0 & 0 & 0 & 0 & 0 \end{bmatrix}$$

The labor force ratio of the primary, secondary, and tertiary industry is 42.5%:33.3%:24.2%; the employment proportions of the four levels in the tertiary industry are as follows: the first level is 11.76%, the second level 2.18%, the third level 4.08%, and the fourth level 6.18%. Then the consumption coefficient matrix T is given by

$$T = \begin{bmatrix} 0.1361 & 0.0133 & 0.0323 & 0.0091 & 0.0354 & 0.0833 \\ 0.2906 & 0.0285 & 0.0689 & 0.0193 & 0.0756 & 0.1779 \\ 0.0518 & 0.0051 & 0.0123 & 0.0035 & 0.0135 & 0.0317 \\ 0.1031 & 0.0101 & 0.0245 & 0.0069 & 0.0268 & 0.0631 \\ 0.1151 & 0.0113 & 0.0273 & 0.0077 & 0.0299 & 0.0705 \\ 0.0699 & 0.0069 & 0.0166 & 0.0047 & 0.0182 & 0.0428 \end{bmatrix}$$

Through matrix A, B, T, and the Leontief "fast track" model, we get the formula:

$$\frac{1}{\alpha}x(t) = Hx(t)$$

Then, $\alpha = 18.6\%$. The unit eigenvalue vector of matrix H is as follows:

(5.71%, 71.20%, 11.006%, 6.92%, 3.35%, 1.82%)

Table 10.23 Industrial Structure of Each Sector in S Province Calculated in Accordance with the Total Social Output Value (%)

YEAR	PRIMARY INDUSTRY	SECONDARY INDUSTRY	FIRST LEVEL	SECOND LEVEL	THIRD LEVEL	FOURTH LEVEL
1998	8.73	72.96	8.41	5.55	2.68	1.67
1999	8.27	72.69	8.70	5.76	2.80	1.69
2000	7.74	72.44	9.20	5.98	2.92	1.72
2001	7.33	72.15	9.58	6.18	3.02	1.74
2002	6.94	71.89	9.92	6.37	3.12	1.76
2003	6.51	71.66	10.30	6.55	3.20	1.78
2004	6.10	71.42	10.65	6.74	3.29	1.80
2005	5.71	71.20	11.00	6.92	3.35	1.82

The actual structure of sectors in 1997 is given by

$$(9.17\%, 73.24\%, 8.06\%, 5.34\%, 2.55\%, 1.64\%)$$

With the use of linear dynamic programming model, the industrial structure of all levels in the primary, secondary, and tertiary industries can be calculated in accordance with the total social output value. Taking the actual industrial structure of each sector in 1997 as the initial conditions for the linear dynamic programming model, the unit eigenvalue vector of the matrix H, namely, the optimal industrial structure as the final constraints, we obtain the industrial structure of each sector from 1998 to 2005 (shown in Table 10.23).

We make the average value of the ratio of the GDP of each industry and the total social output value from 1994 to 2000 as the conversion factor; then the conversion factors in the primary industry, secondary industry, and the four levels of the tertiary industry are 0.4974, 0.2132, 0.5593, 0.6138, 0.5697, and 0.3708, respectively. The internal industrial structure adjustment program of the tertiary industry is shown in Tables 10.24 and 10.25.

According to the input–output table of 1997, we can forecast that the economy of S Province will drive on the "fast track" in 2005. The growth rate of the "fast track" in the tertiary industry is 18%. The ratio of the industrial structure of the primary, secondary, and tertiary industries will reach 9.5:48.3:42.2. In the tertiary industry, the ratio of the output of the four levels is 19.7:13.7:6.2:2.2, the GDP

Table 10.24 Comparison between the Analog Value and the Actual Value of the Structure (%)

	PRIMARY INDUSTRY		SECONDARY INDUSTRY		FIRST LEVEL		SECOND LEVEL		THIRD LEVEL		FOURTH LEVEL	
YEAR	ANALOG VALUE	ACTUAL VALUE	ANALOG VALUE	ANALOG VALUE	ACTUAL VALUE	ANALOG VALUE	ACTUAL VALUE	ANALOG VALUE	ACTUAL VALUE	ACTUAL VALUE	ANALOG VALUE	ACTUAL VALUE
1998	14.0	14.1	51.5	14.0	14.1	51.5	50.6	34.5	35.3	50.6	34.5	35.3
1999	13.6	13.0	51.0	13.6	13.0	51.0	50.9	35.4	36.1	50.9	35.4	36.1
2000	12.7	12.0	50.4	12.7	12.0	50.4	51.6	36.9	36.4	51.6	36.9	36.4

Table 10.25 Internal Restructuring Program in the Tertiary Industry in S Province (%)

YEAR	PRIMARY INDUSTRY	SECONDARY INDUSTRY	FIRST LEVEL	SECOND LEVEL	THIRD LEVEL	FOURTH LEVEL
2001	11.9	50.4	17.5	12.4	5.6	2.1
2002	11.3	50.0	18.1	12.8	5.8	2.1
2003	10.5	49.6	18.7	13.1	5.9	2.1
2004	9.8	49.3	19.3	13.4	6.1	2.2
2005	9.2	49.0	19.7	13.7	6.2	2.2

proportion of the first, second, and third levels rise gradually; only the fourth level remains unchanged. This shows that the pace of development in the tertiary industry is faster than that of the national economy.

11

Research on Approaches to Industrial Structure Upgrading of S Province

11.1 Background of Industrial Structure Upgrading of S Province

Industrial structure upgrading denotes not only accelerating the transformation of traditional industries, but also advancing the modern industries, especially high-tech industries. It should benefit both industrialization and urbanization, adjust urban and rural structures, and reasonably transfer rural surplus labor. As the implementation of the above process cannot be separated from the environment and conditions of the industrial structure system, when we design and practice the upgrading, it is important to systematically analyze and study these aspects.

S Province is one of the developed provinces along the coastal region in China. In 2003, there were 21 counties ranked among 100 strong counties of the whole country. Both economic aggregation and average level of S Province came out at the top in China. In 2003, the GDP of S Province was RMB 1,245,175 million; the per capita GDP reached US$ 2000. According to World Bank standards, the economic development of S Province was roughly equivalent to that of middle-income countries of the mid-1990s. And it is moving into the middle-to-late stage of industrialization from the middle stage. In 2003, the disposable income of the town residents and the gross income of farmers was, 9262 RMB and 4239 RMB, respectively. The quality of life of the residents has significantly improved, housing area has greatly increased, and many high grade consumer durable goods are being purchased by common people.

The upgrading of industrial structure cannot be maintained without the support of science. By 2002, there were 1890 scientific institutions in S Province and 2839 in 2003, with 109,400 employees. Among these institutions, 348 institutions were independent, 588 institutions were affiliated to universities, and there were also 941 large and medium enterprise research institutions. S Province had built 30 national and provincial level laboratories, 92 technical project research centers, 11 public technical service platforms, 19 national level software parks and high-tech industry bases. In 2003, there were 2173 high-tech enterprises in S Province. The high-tech industry output achieved RMB 382.796 billion and increased by 51.44% compared with 2002. There were also 14 national and provincial level high-tech industrial parks, which achieved an output of RMB 400 billion. It had 25 national level high-tech industrial bases for new materials, software, sensors, and new drugs. It ranked first in China in the quantity.

In recent years, the economy has progressed rapidly in S Province. There has been more and more positive factors such as urban expansion, increasing foreign and private capital investment, and system innovation. The foundation of rapid economic growth is strengthening, which provides basic condition and material base for upgrading industrial structure. At the same time, there are also some problems which cannot be ignored.

1. The pressure of employment and reemployment is still high. Although the economy is growing fast, the urban unemployment rate is quite high. On the one hand, the number of persons laid-off from enterprise reform is higher than ever, which indicates that the chance of reemployment of persons laid-off is less. On the other hand, there is an excess supply of labor force. The new labor force in cities and the transfer of rural labor increases the problem of employment. However, the push of economic growth for employment has become weaker in recent years, as the employment elasticity has decreased.
2. The increasing income disparity affects consumption. The income disparity between the urban and rural regions is increasing. In 1990, the ratio of urban disposable income and per capita gross income farmers was 1.66:1. In 2001, the ratio was 1.95:1 and 2.05:1 in 2002. Second, the income disparity of unban residents is also increasing. In cities, the income of 60% of residents is below the average level of the province; 10% of the high-income families get 10 times more than 10% of the

low-income familes. In rural areas, the income of 60% of the people is below average level. The disparity between 10% of the high-income and 10% low-income families is about eight times. Third, the balance of payments expectation of most residents is not stable when facing house, medical, education, and endowment reform. These vulnerable groups have high expectation for future expenditure, so that they would rather save than spend money. As a result, consumption was inhibited.

3. Regional economic development is not balanced. After many years of hard work, the strategy of common development in the whole province resulted in great achievement. However, there is still a great gap between the southern, central, and northern regions in various factors. Taking the speed of economic growth as an example, the total industrial value of the five cities in the southern region makes up 70% of the whole province, that is, an increase of 16%. And it has a contribution of 73% for the growth of industry, which makes a growth of 11.2% to industry. As to the central and northern regions, the industrial value added increased to 13% and 15.3%, the contribution to industry were 13.7% and 13.3%, respectively. The southern region accounts for 88.5% of the FDI, the central and northern regions account for 6.3% and 5.2%, respectively. This has a bad effect on resource allocation and consolidated strength of the whole province and implies that regional economic development is not balanced in the whole province.
4. There are still some potential problems in the operation of microeconomy. First, the benefit for state-owned enterprises is low. In 2003, 36.1% of state-owned industrial enterprises ran at a loss. Especially, there were 246 among 750 large and middle enterprises that were running at a loss. Second, the foreign funded processing type enterprises play an important role in economic growth; even they are easily affected by the global economic wave. Among all the scale-sized enterprises, foreign funded enterprises make 30% of industrial value added. Further, foreign funded enterprises account for 60% of exports. The fluctuation of the foreign funded enterprises will have great influence on the economic growth of the whole province if the domestic and international environment

changes. Third, real estate enterprises have problems in their operation. Though the rapid growth of real-estate investment is helpful to economic growth, people are still worried about the increasing price of houses and high loan and high-debt operations of some real-estate enterprises. According to a survey of 183 real-estate enterprises, in 2002, their own fund accounted only for 18.6%. We can expect that the average asset–liability ratio will be 85.7%, and the profit rate of operation and that of cost will be decreasing.

Given such a background, we should work according to the principles of scientific development, walk the road with new industrialization and fine overall arrangements. We have much to do to upgrade the industrial structure of S Province.

11.2 Comprehensive Evaluation and Analysis of the Industrial Structure of S Province

11.2.1 Association Analysis of the Industrial Structure of S Province

In the macroeconomy, there are different relations between industries, such as the relation of technology and economy and the relation of nature and society. Therefore, the connotation of industrial association includes direct, indirect, and derivational influence among industries. The index to measure the association of industrial input–output system is known as correlation coefficient, including the sensitivity coefficient μ_i and the influence coefficient γ_j. These indexes reflect the technique of economic relation of all sectors of national economy in production, exchange, and distribution.

All industrial sectors can be divided into four types depending on their sensitivity coefficient and influence coefficient. (1) Dull association ($\mu_i < 1$, $\gamma_j < 1$): this type of industry has loose relationship with other industries; (2) inductive association ($\mu_i > 1$, $\gamma_j < 1$): this type of industry has strong sensitivity to other industries and mainly "pushes" other industries to develop; (3) influence association ($\mu_i < 1$, $\gamma_j > 1$): this type of industry usually "pulls" other industries to develop; and (4) sensitive association ($\mu_i > 1$, $\gamma_j > 1$): these industries usually "push" and "pull" the economic development at the same time. Usually, these are chosen as leading industries.

11.2.1.1 Comparison of Input–Output Correlation Coefficients There were 17 sectors in the input–output tables of S Province in 2000, but in 1997, there were 40 sectors. We combine some sectors of the 1997 input–output tables according to the standard of 2000 and calculate the sensitivity coefficient and influence coefficient of all industries (see Table 11.1).

All 17 sectors can be divided as follows:

Dull association: Agriculture, electricity, steam and water production, and supply, finance and insurance, social service, other services

Inductive association: Extractive industry, commercial catering

Table 11.1 Input–Output Correlation Coefficient of S Province

	1997		2000	
INDUSTRY	SENSITIVITY COEFFICIENT	INFLUENCE COEFFICIENT	SENSITIVITY COEFFICIENT	INFLUENCE COEFFICIENT
Agriculture	0.7969 (8)	0.8101 (15)	0.7897 (8)	0.8218 (14)
Extractive industry	1.0031 (4)	0.8230 (14)	1.1018 (4)	0.8119 (15)
Food manufacturing and tobacco processing	0.7430 (11)	1.0591 (8)	0.7277 (12)	1.0876 (7)
Textiles, clothing, leather, and other fiber manufacturing	0.8347 (7)	1.1724 (5)	0.8059 (7)	1.1792 (5)
Coking, gas, and petroleum processing	2.0123 (2)	1.1010 (7)	2.1593 (2)	1.0750 (8)
Metal manufacturing	0.6555 (14)	1.1849 (4)	0.6346 (15)	1.1820 (4)
Machinery	0.7358 (13)	1.1902 (3)	0.6864 (13)	1.2035 (3)
Electronic and communication equipment manufacturing	0.7706 (10)	1.2003 (2)	0.7657 (10)	1.2168 (1)
Other manufacturing	3.1823 (1)	1.2037 (1)	3.0661 (1)	1.2049 (2)
Electricity, steam and water production, and supply	0.8411 (6)	0.8583 (13)	0.8414 (6)	0.8702 (11)
Architecture	0.4953 (16)	1.1174 (6)	0.4671 (16)	1.1227 (6)
Transportation, storage, and post and telecommunications	0.9899 (5)	0.7909 (16)	1.0053 (5)	0.7951 (16)
Commercial catering	1.3788 (3)	0.8718 (12)	1.3229 (3)	0.8637 (12)
Finance and insurance	0.7412 (12)	0.6973 (17)	0.7418 (11)	0.7131 (17)
Real estate	0.4386 (17)	1.0455 (9)	0.4577 (17)	1.0268 (9)
Social services	0.7739 (9)	0.8919 (11)	0.7879 (9)	0.8502 (13)
Other services	0.6064 (15)	0.9823 (10)	0.6387 (14)	0.9755 (10)

Source: Input–output table of 1997 and 2000 in S Province.

Note: The numbers in the parentheses indicate the order of the calculated correlation coefficients in corresponding column.

Influence association: Food manufacturing and tobacco processing industry, textiles, clothing, leather and other fiber manufacturing industry, metal manufacturing, machinery, electronic and communication equipment manufacturing industry, architecture, real estate industry

Sensitive association: Coking, gas and petroleum processing industry, other manufacturing industry

Transportation, storage, and posts and telecommunications industry were classified under dull association in 1997. However, as the sensitivity coefficient increased in 2000, they were grouped under inductive type association industries.

11.2.1.2 Comparison of Input–Output Tables of 1997 and 2000

11.2.1.2.1 Comparison of Correlation Coefficients The correlation coefficients of all sectors changes little between 1997 and 2000. The sensitivity coefficients of top 10 industries remained the same. They are the other manufacturing industry, coking, gas and petroleum processing industry, commercial catering, extractive industry, transportation, storage, posts and telecommunications industry, electricity, steam and water production, and supply industry, textiles, clothing, leather and other fiber manufacturing industry, agriculture, social service industry, electronic and communication equipment manufacturing industry. Food manufacturing and tobacco processing industry ranked 11 in 1997 and 12 in 2000, finance and insurance ranked 12 in 1997 and 11 in 2000. The machinery industry ranked 14 in 1997 and 15 in 2000, other service industry ranked 15 in 1997 and 14 in 2000.

The influence coefficient of top 10 industries also remained the same in 1997 and 2000, including other manufacturing industry, machinery industry, metal manufacturing industry, textiles, clothing, leather and other fiber manufacturing industry, architecture, coking, gas and petroleum processing industry, food manufacturing and tobacco processing industry, real estate industry, other service industry. Other manufacturing industry ranked 1 in 1997 and 2 in 2000, the electronic and communication equipment manufacturing industry ranked 2 in 1997 and 1 in 2000. Coking, gas and petroleum processing industry ranked 7 in 1997 and 8 in 2000, food manufacturing and tobacco processing industry ranked 8 in 1997 and 7 in 2000.

Finance and insurance ranked in 11 in 1997 and 13 in 2000, machinery industry ranked 13 in 1997 and 11 in 2000. Extractive industry ranked 14 in 1997 and 15 in 2000, agriculture ranked 15 in 1997 and 14 in 2000. These changes are the results of industry technology development and industrial structure adjusting policies.

11.2.1.2.2 Comparison of Total Amount In the input–output table of 1997, the total flow of intermediate goods was RMB 145.08197 billion. The number reached RMB 184.085579 billion in 2000; an increase of 26.88%. It shows that the relation of all industries of S Province increased the quantity of flow of goods.

The total output of 1997 was RMB 210.068996 billion, and intermediate goods accounted for 69.06%. The total output of 2000 was RMB 268.730642 billion and intermediate goods accounted for 68.50%. From the percentage value of intermediate goods, it is seen that the economic and technical relations of all industries decrease.

11.2.1.2.3 Evolution of Direct Input Coefficient The direct input coefficient denotes the ratio of all intermediate goods consumption to total output of this industry. Table 11.2 shows the evolution of direct input coefficient from 1997 to 2000.

In the input–output tables of 1997 and 2000, the industries whose direct input coefficient are above 0.7 include food manufacturing and tobacco processing industry, textiles, clothing, leather and other fiber manufacturing industry, coking, gas and petroleum processing industry, metal manufacturing industry, machinery industry, electronic and communication equipment manufacturing industry, other manufacturing industry and architecture. These industries consume much intermediate goods during production. In other words, the relation between other industries is stronger in the economic system.

In the input–output tables of 1997 and 2000, the industries whose direct input coefficients are between 0.5 and 0.7 include electricity, steam and water production, and supply industry, commercial catering, real estate industry, social service industry and other service industries. In the production process, the main input of these industries are intermediate goods.

In the input–output tables of 1997 and 2000, the industries whose direct input coefficients are less than 0.5 include agriculture,

Table 11.2 Evolution of Direct Input Coefficient

INDUSTRY	DIRECT INPUT COEFFICIENT 1997	2000	RATIO OF CHANGE (%)
Agriculture	0.4995	0.497	−0.501
Extractive industry	0.5042	0.493	−2.221
Food manufacturing and tobacco processing	0.765	0.7744	1.229
Textiles, clothing, leather, and other fiber manufacturing	0.7655	0.7639	−0.209
Coking, gas, and petroleum processing	0.7526	0.7294	−3.083
Metal manufacturing	0.7715	0.763	−1.102
Machinery	0.7642	0.769	0.628
Electronic and communication equipment manufacturing	0.7688	0.7714	0.338
Other manufacturing industry	0.7932	0.7879	−0.668
Electricity, steam and water production, and supply	0.5865	0.5928	1.074
Architecture	0.7179	0.7173	−0.084
Transportation, storage, and post and telecommunications	0.4494	0.4447	−1.046
Commercial catering	0.6105	0.5901	−3.342
Finance and insurance	0.3759	0.3864	2.793
Real estate	0.6673	0.6454	−3.282
Social services	0.565	0.5192	−8.106
Other service industry	0.6142	0.6034	−1.758

Source: Input–output tables of 1997 and 2000 in S Province.

transportation, storage, posts and telecommunications industry, finance and insurance industry. In the production process, the value added ratio is higher in these industries.

Further, the direct input coefficient of extractive industry is 0.5042 in 1997 and 0.493 in 2000. Overall, the consumption of intermediate goods and initial input of these industries were almost the same.

From single industry perspective, the industries whose direct input coefficients decreased are agriculture, extractive industry, textiles, clothing, leather and other fiber manufacturing industry, coking, gas and petroleum processing industry, metal manufacturing industry, other manufacturing industry, architecture industry, transportation, storage, posts and telecommunications industry, commercial catering, real-estate industry, social service industry and other service industry. The four industries whose direct input coefficient decreased most were the social-service industry, coking, gas and petroleum processing industry, commercial catering, and real-estate industry. The extent of reduction was by 8.106%, 3.083%, 3.342%, and 3.282% respectively.

The industries whose direct input coefficient increased are food manufacturing and tobacco processing industry, machinery industry, electronic and communication manufacturing industry, electricity, steam and water production, and supply industry, and finance insurance industry. The reductions were by 1.229%, 0.628%, 0.338%, 1.0742%, and 2.793% respectively.

If we evaluate the inner correlation of the economic system according to the sum of the direct input coefficients, the value in 1997 was 10.971, and became 10.848 in 2000, a decrease by 1.121%. The correlation reduces a little, which was because of fluctuation during development.

11.2.1.2.4 Evolution of Intermediate Demand Coefficient The intermediate demand coefficient denotes the ratio of intermediate demand and total output. The value is usually less than 1, which denotes that the output of one industry meets the requirement of direct input. If the value is more than 1, it denotes that the output does not meet the requirement. The product of that industry should be imported from other regions. Table 11.3 shows the evolution of intermediate demand coefficient from 1997 to 2000.

The intermediate demand coefficient of extractive industry, transportation, storage, posts and telecommunications industry are more than 1, which indicates that these two industries cannot meet the intermediate demand of the economic system. The net products of these industries should be imported.

In all 17 industries, the intermediate demand coefficient of 13 industries declined, including extractive industry, food manufacturing and tobacco processing industry, textiles, clothing, leather and other fiber manufacturing industry, metal manufacturing industry, machinery industry, electricity, steam and water production, and supply industry, architecture industry, transportation, storage, posts and telecommunications industry, commercial catering, finance and insurance industry, social-service industry. It increased in the case of four industries, including agriculture, coking, gas and petroleum processing industry, real-estate industry, other service industry.

Table 11.3 Evolution of Intermediate Demand Coefficient

INDUSTRY	INTERMEDIATE DEMAND COEFFICIENT 1997	INTERMEDIATE DEMAND COEFFICIENT 2000	RATIO OF CHANGE (%)
Agriculture	0.463709	0.471223	1.621
Extractive industry	3.025781	2.765552	–8.600
Food manufacturing and tobacco processing	0.52512	0.519638	–1.044
Textiles, clothing, leather, and other fiber manufacturing	0.541436	0.519413	–4.067
Coking, gas, and petroleum processing	0.903382	0.923318	2.207
Metal manufacturing	0.759557	0.754663	–0.644
Machinery	0.447251	0.409113	–8.527
Electronic and communication equipment manufacturing	0.656776	0.630566	–3.991
Other manufacturing	0.897832	0.884973	–1.432
Electricity, steam and water production, and supply	0.964562	0.952497	–1.251
Architecture	0.093293	0.060407	–35.250
Transportation, storage, and post and telecommunications	1.109263	1.075583	–3.036
Commercial catering	0.866171	0.835711	–3.517
Finance and insurance	0.854569	0.853679	–0.104
Real estate	0.287012	0.325303	13.341
Social services	0.756976	0.748988	–1.055
Other service industry	0.27902	0.287195	2.929

Source: Input–output tables of 1997 and 2000 in S Province.

If we evaluate the correlation of the economic system according to the sum of intermediate demand coefficient, the value in 1997 was 13.432, in 2000 it became 13.019. The value decreased by 3.075%. The trend is the same as direct input coefficient.

11.2.2 Quadratic Programming Mathematical Model of Industrial Structure of S Province

Like the developing law of everything, evaluation of industrial structure is from quantitative change to qualitative change, which has different stages. So is case with S Province.

The industrial structure has changed a lot from 1978 to 2002. Figure 11.1 shows the graph of industrial structure of S Province from 1978 to 2002. The GDP of secondary industry remains at 50%. The percentage of primary industry declines from 27.57% to 10.53%, and that of the tertiary industry increases from 19.83% to 37.3%

We will study the evolution trend of GDP structure base on the GDP of industries of S Province from 1978 to 2002. We will establish a quadratic programming model based on system structure transfer homogeneous Markov chain. The future output depends on the present situation and not on the past, so we presume that the transfer progress has the properties of homogeneous Markov chain.

X is the GDP of three industries of S Province; $X_i(t)$ is the GDP of industry i at time t; $W_i(t)$ is the ratio of GDP of industry i at time t to the total GDP, $X_i(t)/X$; P_{ij} is the one-step transfer probability of GDP of industry i at time $(t-1)$ transfers at time t;

$$W_j(t) = \sum_{i=1}^{3} W_i(t-1)P_{ij} + \varepsilon_j(t)$$

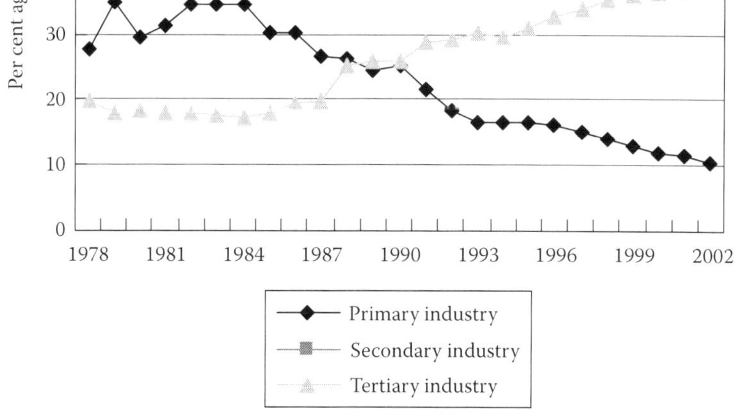

Figure 11.1 Graph of industrial structure of S Province from 1978 to 2002.

The P_{ij} we get should have the least error $\sum_{t=1}^{n-1}\sum_{j=1}^{3}\varepsilon_j(t)$. We establish the quadratic programming model as follows:

$$\min \sum_t \sum_j \left(W_j(t) - \sum_{i=1}^{3} W_i(t-1)P_{ij} \right)^2 \quad \text{s.t.} \begin{cases} \sum_{i=1}^{3} P_{ij} = 1 \\ P_{ij} \geq 0 \end{cases}$$

We divide the data into two parts; one from 1978 to 1992, the other from 1993 to 2002. We solve the quadratic programming using MATLAB. The transition probability matrices are as follows:

$$P_{(1978-1992)} = \begin{bmatrix} 0.962752 & 0.002054 & 0.035194 \\ 0 & 0.999537 & 0.000463 \\ 0 & 0 & 1 \end{bmatrix}$$

$$P_{(1993-2002)} = \begin{bmatrix} 0.932409 & 0 & 0.067591 \\ 0 & 0.996176 & 0.003824 \\ 0 & 0 & 1 \end{bmatrix}$$

We easily find that from 1978 to 1992, the speed of transfer from the primary industry to the tertiary industry is quite high. People who transfer from the primary industry to the secondary and tertiary account for 0.21% and 3.52%. The percentage of GDP which is transfered from the secondary to the tertiary industry is 0.05%.

The GDP structure of S Province from 1992 to 2002 has also changed. The primary industry stops transfer to the secondary industry, and speeds up transfer to the tertiary industry, the percentage value of change is from 3.52% to 6.76%. The secondary industry also accelerates transferring to the tertiary industry, from 0.05% to 0.38%. Briefly speaking, both the primary and secondary industries transfer to the tertiary industry from 1992 to 2002.

The GDP of S Province comes out top in China. The change of three industries shows the transfer of productivity. Industry (the secondary industry) plays an important role in S Province, and its development is always smooth. We see from the above transfer matrix that during the two phases, agriculture productivity transfers to the secondary and tertiary industry. Although it mainly transfers to the tertiary industry, the emphasis point changes. During the first phase (1978–1992), the primary industry mainly transfers the productivity to the tertiary industry and a little to the secondary industry. During the second phase, it stops transfer to the secondary industry and the secondary industry begins the transfer to the tertiary industry. After many years of development, the primary industry has transferred most of the productivity, so it will not be the focus of productivity in the future. From the transfer of secondary industry, this phase can be considered as a transitional stage which has just a little change. The transfer of productivity to tertiary industry is always quick. In the first stage, the productivity of the tertiary industry increases together with that of the secondary industry. Then both the primary and secondary industries transfer to the tertiary industry which makes the tertiary industry develops fast. We think that this trend will not change in a short time, and we predicted that in 2005, the proportion of three industries of S Province will be 8.7%, 50.4%, and 40.9%. The primary industry will decline in future, but the amplitude will reduce because there is not much space. The percentage of the tertiary industry will increase and approach that of the secondary industry.

11.2.3 Comparative Analysis of Industrial Structure between S Province and Other Provinces

Comparing the industrial structure and economic situation of S Province with other similar ones such as Shanghai, Zhejiang, Guangdong, Shandong, and Fujian can give us some idea in finding out the shortcomings of the industrial structure of S Province.

Table 11.4 shows the evolution of industrial structure of different provinces since 1978 and the GDP of every year. In 1985, the per capita GDP of S Province, Zhejiang, and Guangdong were almost the same. However, the per capita GDP of Zhejiang and Guangdong are higher than that of S Province since 1990. Studying the industrial labor structure of different provinces, we find that the change of labor structure of S Province is lower than that of other provinces (except that of Zhejiang from 1985 to 1990). The above situation indicates that, at that time, the adjustment of employment structure occurred

Table 11.4 Evolution of Industrial Structure of Different Provinces

INDUSTRY PROPORTION AND GDP COMPOSITION		1978	1980	1985	1990	1995	1996	1997	1998	1999	2000	2001	2002
S Province	Primary	27.6	29.5	30.0	25.1	16.4	16.1	15.1	14.1	13.0	12.0	11.4	10.5
	Secondary	52.6	52.3	52.1	48.9	52.7	51.2	51.1	50.6	50.9	51.7	51.6	52.2
	Tertiary	19.8	18.2	17.9	26.0	30.9	32.7	33.8	35.3	36.1	36.3	37.0	37.3
	Per capita GDP (RMB)	430	541	1053	2103	7299	8447	9344	10,021	10,665	11773	12,922	14,391
Shanghai	Primary	4.0	3.2	4.2	4.3	2.5	2.5	2.3	2.1	2	1.8	1.7	1.6
	Secondary	77.4	75.7	69.8	63.8	57.3	54.5	52.2	50	48.4	47.5	47.6	47.4
	Tertiary	18.6	21.1	26.1	31.9	40.2	43	45.5	47.8	49.6	50.6	50.7	51
	Per capita GDP (RMB)	2498	2738	3855	5910	18,942	22,275	25,750	28,240	30,805	34,547	37,282	40,646
Zhejiang	Primary	38.1	36.0	29.0	25.1	15.9	14.7	13.7	12.7	11.8	11	10.3	8.8
	Secondary	43.3	46.8	46.5	45.4	52	53.1	54.1	54.3	54.1	52.7	51.3	51.2
	Tertiary	18.6	17.2	24.5	29.5	32.1	32.2	32.2	33	34.1	36.3	38.4	40
	Per capita GDP (RMB)	329.8	469.5	1050	2120	8067	9423	10,488	11,216	12,009	13,410	14,930	16,838
Shandong	Primary	33.29	36.43	34.68	28.14	20.19	20.13	17.97	16.98	15.94	14.85	14.40	13.17
	Secondary	52.94	50.02	43.07	42.09	47.43	47.16	47.90	48.27	48.36	49.96	49.32	50.32
	Tertiary	13.77	13.55	22.25	29.77	32.38	32.71	34.13	34.75	35.70	35.46	36.28	36.51
	Per capita GDP (RMB)	314.9	400.4	884.3	1794	5749	6814	7548	8073	8588	95,183	10,459	11,635
Guangdong	Primary	29.8	33.2	29.8	24.7	15.1	14.4	13.5	12.7	12.1	10.4	9.5	8.8
	Secondary	46.6	41.1	39.8	39.5	50.2	50.2	49.9	50.4	50.4	51.1	50.3	50.2
	Tertiary	23.6	25.7	30.4	35.8	34.7	35.4	36.6	36.9	37.5	38.5	40.2	41
	Per capita GDP (RMB)	366.9	477.6	1021	2496	8446	9453	10430	11,129	11,597	11,802	13,680	14,975

Table 11.5 Output Structure of Faming, Forestry, Animal Husbandry and Fisheries (%)

INDUSTRY	1952	1957	1962	1978	1980	1985	1990	1995	2000	2001
Faming	82.02	81.77	85.16	80.45	76.55	69.55	62.44	58.46	58.62	58.42
Forestry	0.09	0.60	0.80	1.40	1.40	1.60	1.37	1.27	1.61	1.57
Animal husbandry	15.69	14.26	11.16	15.85	19.25	23.06	27.70	28.20	23.03	22.93
Fisheries	2.20	3.37	2.89	2.30	2.80	5.38	8.50	12.07	16.74	17.08

Source: Statistical Yearbook of S Province.

later than in other provinces. This reflects the effect of structural change on economic benefit.

11.2.3.1 Comparison of the Inner Structure of the Primary Industry of Different Provinces With the development of economy, the inner structure of the primary industry has changed a lot. The percentage of agricultural industry has declined and percentage of animal husbandry, especially fisheries, has increased. According to Table 11.5, the inner structure of the primary industry shows no change before 1978, which is because of the industrial policy at that time. Industrial structure changed a lot after reform and opening up of the economy. Comparing the ouput structure of 1978 with that of 2001, agricultural industry decreased by 22.03%, fisheries and animal husbandry increased by 14.62% and 7.08%, respectively.

We divide the primary industry into two parts: agriculture and others. The conditions in different provinces in 2001 is shown in Table 11.6.

The study on the inner structure of primary industries of China and S Province shows that the percentage of agricultural industry has

Table 11.6 Comparison of Inner Structure of Agriculture Industry of Some Provinces in 2001 (%)

	INDUSTRY	
PROVINCE	AGRICULTURE	OTHERS
China	55.24	44.76
S Province	58.42	41.58
Zhejiang	47.79	52.21
Shanghai	41.97	58.03
Guangdong	48.46	51.54
Fujian	40.81	59.19
Shandong	57.11	42.89

Source: China Statistical Yearbook of 2002.

declined and that of the other industries has increased. The percentage of agricultural industry of S Province is 3.18% more than that of China, 10% more than Zhejiang, Guangdong, Fujiana, and Shanghai. This indicates that S Province was behind the country in the adjustment of primary industry more than that of other provinces.

We divide the agricultural industry into two parts: food crops and economic crops. The ratio of agricultural area of all provinces in 2001 is shown in Table 11.7.

The proportion of economic crops of S Province is 5.3% more than the average level in China, which indicates the adjustment of food and economic crops of S Province is at the top place in the country. However, it is behind Zhejiang, Guangdong, and Shanghai. The proportion of economic crops is 3.09% and 2.64% lower than that of Zhejiang and Guangdong, repsectively.

11.2.3.2 Comparison of Inner Structure of Secondary Industry of Different Provinces We divide the secondary industry into industry and construction and study the proportion of GDP of all provinces.

From Table 11.8 we see that the ratio of industry of S Province is just below that of Zhejiang, and the ratio of construction and total GDP is top in the country. Briefly, the percentage of the secondary industry is higher than that of other provinces. Industry and construction industry play an important role in the economic system in S Province.

We see from Table 11.9 that the total output of S Province in 2002 was 1,386,586 million ¥. The ratio to Shanghai, Shandong, and Zhejiang are 179.1%, 120.6%, and 175.9%, respectively. The heavy

Table 11.7 Areas of Crops in Provinces in 2001

PROVINCE	TOTAL AREA UNDER CROPS (1000 ha)	FOOD CROPS (%)	ECONOMIC CROPS (%)
China	155,708	68.13	31.87
S Province	7,777.4	62.83	37.17
Zhejiang	3,245.9	59.74	40.26
Shanghai	490.9	43.02	56.98
Guangdong	5,193.1	60.19	39.81
Fujian	2,713.1	63.61	36.39
Shandong	11,266.1	63.50	36.50

Source: China Statistical Yearbook of 2002.

Table 11.8 Proportion of Industry and Construction in 2001

	GDP (BILLION RMB)	PROPORTION OF SECONDARY INDUSTRY (%)	INDUSTRY (%)	CONSTRUCTION (%)
S Province	951.191	51.59	44.90	6.69
Shanghai	495.084	47.58	42.85	4.73
Zhejiang	674.815	51.27	46.03	5.24
Shandong	943.831	49.32	43.36	5.96
Guangdong	1064.771	50.17	44.45	5.72
Fujian	425.368	44.77	38.68	6.09

Source: China Statistical Yearbook of 2002.

Table 11.9 Comparison of the Inner Structure of Secondary Industries of Some Provinces

	CHINA (2001)	S PROVINCE (2002)	SHANGHAI (2002)	SHANDONG (2002)	ZHEJIANG (2001)
Gross industrial output (billion RMB)	9544.898	1,386.586	774.056	1,149.753	788.247
Light industry	39.43	42.20	33.89	44.66	55.46
Heavy industry	60.57	57.80	66.11	55.34	44.54
Extractive industries	5.66	1.03	0.37	8.35	0.38
Food manufacturing and tobacco processing	9.69	6.13	0.95	15.31	6.17
Textiles and clothing	10.26	16.41	7.13	10.53	24.40
Wood and paper industry	4.6	4.53	2.61	4.79	6.20
Petroleum refining chemical and pharmaceutical industry	17.8	18.98	17.21	17.61	17.67
Nonmetallic mineral products industry	4.22	3.56	2.26	5.45	3.18
Metallurgy and metal manufacturing	11.45	11.28	12.92	6.90	7.95
Machinery	6.14	10.11	8.44	9.30	9.03
Transport equipment	6.78	5.40	14.42	4.72	4.97
Electrical, electronic instrument, and meter manufacturing	16.14	17.97	22.67	10.81	13.67
Other manufacturing industries		1.07	0.71	1.45	
Public utilities	5.88	3.51	3.34	4.73	4.23

Source: Calculated according to *Statistical Yearbook* of different provinces.

industry accounts for 57.80%, which is 2.46% more than that of Shandong, 10.8% more than Zhejiang, 8.31% less than that of Shanghai and 2.77% less than the average level of the country. These numbers denote that the industrialization process of S Province was in leading place.

As to the inner structure of the secondary industry, the top five output industries of S Province are petroleum refining and chemical industry, pharmaceutical industry (18.98%), electrical and electronic instruments and meter manufacturing industry (17.97%), textiles and clothing industry (16.41%), metallurgy and metal manufacturing industry (11.28%), machinery industry (10.11%). The total percentage of these five industries is 74.57%. Four of them belong to heavy industry except textiles and clothing industry. Chemical and pharmaceutical industry, electrical and electronic instruments and meter manufacturing, machinery parts industry are all high-tech industries, which indicate that S Province has advantage in this domain.

The top five output industries of Shanghai are electrical and electronic instruments and meter manufacturing industry (22.67%), petroleum refining and chemical industry, pharmaceutical industry (17.21%), transportation equipment manufacturing industry (14.42%), metallurgy and metal manufacturing industry (12.92%), machinery industry (8.44%). All the five industries are heavy industries and their total output accounts for 75.66%. Electrical and electronic instruments and meter manufacturing industry of Shanghai is 4.7% to 11.86% more than that of other provinces, transportation equipment manufacturing industry is about 10% more than that of other provinces, which are obvious characteristics.

The top five output industries of Zhejiang are textiles and clothing industry (24.40%), petroleum refining and chemical industry, pharmaceutical industry (17.67%), electrical and electronic instruments and meter manufacturing industry (13.67%), machinery industry (9.03%), metallurgy and metal manufacturing industry (7.95%). The total percentage is 72.72%. These industries are the same as those of S Province, which indicates that the industrial structure of the two provinces is similar. As to the percentage of every industry in the GDP, textiles and clothing industry plays an important role in Zhejiang industrial structure, which accounts for one quarter of the total output.

The important industries of Shandong are petroleum refining and chemical industry, pharmaceutical industry (17.61%), food manufacturing and tobacco processing industry (15.31%), electrical and electronic instruments and meter manufacturing industry (10.81%), textiles and clothing industry (10.53%), machinery industry (9.30%). The total percentage is 63.56%. We find that food manufacturing and tobacco processing industry and textiles and clothing industry are developed. This is similar to S Province.

11.2.3.3 Comparison of the Inner Structure of Tertiary Industry of Different Provinces Since 1978, the tertiary industry of S Province is developing fast, the average increase is 15%, which is faster than the development of GDP. If we study the structure, we note that the proportion of output and labor of tertiary industry is lower than that of other industries.

Table 11.10 shows that the commercial catering industry is quite important in the tertiary industry. Its output accounts for one quarter. The finance insurance industry is developed in Shanghai, and its output also accounts for about one quarter; this is also the feature of a financial center. The inner structure of the tertiary industry of Shandong is similar to that of S Province. Transportation, posts and

Table 11.10 Comparison of the Inner Structure of Tertiary Industry of Some Provinces in 2001

	S PROVINCE	SHANGHAI	ZHEJIANG	SHANDONG	GUANGDONG	FUJIAN
Tertiary industry (RMB 100 million)	3522.02	2509.81	2593.25	3424.31	4301.75	1698.36
Post and telecommunication (%)	18.31	13.74	19.42	19.52	24.96	27.58
Commercial catering (%)	27.18	21.93	35.05	26.03	24.28	24.23
Finance and insurance (%)	12.79	24.7	9.41	13.59	8.58	11.87
Real estate (%)	11.18	12.62	5.64	10.99	13.01	7.57
Social services (%)	8.77	10.84	10.7	6.37	14.32	10.57
Science, education (%)	11.12	12.02	12.69	12.19	8.25	10.31
Others (%)	10.71	4.15	7.09	11.31	6.6	7.86

Source: China Statistical Yearbook of 2002.

telecommunications industry and commercial catering industry play important roles in the tertiary industry of Guangdong and Fujian; the ratio of the output are almost the same.

11.3 Approaches to Industrial Structure Upgrading of S Province

There are two ways to upgrade the industrial structure. One is changing the structure of different industries: at first, the primary industry is the leading one, then the secondary industry, and finally, the tertiary industry. Relatively, the leading industry changes from being labor intensive to capital intensive, then to technology intensive. The second method is the upgrading within the industries. It mainly denotes the development of technology, including the replacement of different technologies and products. Both ways are manifestations of industrial structure situation and the upgrading ability, which is the key to developing the economy. The two ways exist simultaneously during the upgrading of industrial structure. However, the content and emphasis points vary in different regions and different developing conditions.

No matter which way is taken, realizing industrial structure upgrading needs many specific approaches and relevant policies. In upgrading the industrial structure of S Province, it is not feasible to imitate everything other countries have done. Therefore, it is important to design a suitable way depending on the situation of S Province.

11.3.1 General Targets of Upgrading the Industrial Structure of S Province

We should follow some rules in upgrading the industrial structure. In the case of S Province they are as follows: balancing urban and rural development, regional development as a whole, development of an economic society, properly balancing harmonious development of humans and nature, balancing domestic development and opening up to the outside world. We should also stick to the position of human-oriented scientific development, keep the development of national economy sustainable, fast, harmonious, and healthy; stick to drive industrialization through information and to accelerate information by industrialization, to walk a high-tech, high benefit, low resource consumption, low pollution, high efficiency and new industrialization way; stick to drive the upgrading of industrial structure by a combination of economy and practicality, and grasping the key point of industrial structural adjustment. Then an industrial pattern will gradually transform to such a developing pattern where agriculture is the foundation, high-tech industry is leading, basic industries and manufacturing industries are supporting and service industries developing completely. This can also improve the competiveness of the whole industry system. With enterprises as the main body, we should also form a new mechanism, on the foundation of full-playing market, for resource distribution, combining the adjustment of the market and the government. Further, positively pushing a reasonable industrial division of S Province and other regions (home and abroad) is also required.

The general targets of upgrading are extending the industrial chain; improving quality of products; opening up the market and increasing effective supply by developing industries which have high demand elasticity, low consumption, and high benefit; keeping a healthy growth of economy by increasing the technology research, and producing more new products by labor intensive industries which can widen the domain of employment and feasibly adjust the industrial structure. We work out the best industrial structure for S Province according to the practice of S Province and the turnpike model. Then we work out the best plan to adjust the industrial structure during the 11th Five-Year Plan by using the dynamic linear programming model (Table 11.11).

We see that the target of upgrading the industrial structure is as follows: By 2010, the ratio of the three industries is 7.66:49.01:43.33. The added value of high-tech industries can significantly increase. We should cultivate some industries which have comparative advantage

Table 11.11 Plan of Industrial Structure of S Province According to GDP

	PRIMARY INDUSTRY	SECONDARY INDUSTRY	TERTIARY INDUSTRY
2004	0.0969	0.5095	0.3937
2005	0.0932	0.5065	0.4003
2006	0.0897	0.5034	0.4069
2007	0.0862	0.5002	0.4135
2008	0.0829	0.4969	0.4201
2009	0.0797	0.4935	0.4268
2010	0.0766	0.4901	0.4333

and competitive advantage to becoming the leading industries because of the fierce competition at home and abroad. The technology of traditional industries would be enhanced and large enterprise groups would be formed. We can establish a new innovation system in which the enterprises are the main body. The main targets are as follows:

- The industrial structure should be diversified and upgraded. We should positively cultivate technical, high value-added industries and high-tech industries on the basis of reform, optimization, and consolidation and enlarge import of leading products which have comparative advantage. This can make the industrial structure suitable for multiple demands.
- The organizational structure should be hierarchical. We will practice the strategy of large enterprise groups with products and capital flow. Then few large enterprises and enterprise groups will be the leading ones in quality of products, process levels, equipment, and efficiency. They will have the ability for advanced development. Some large enterprises and enterprise groups will have high technology, can produce high value-added goods and be competitive at home and abroad. Most enterprises should continuously improve their technology, create new products, and participate in the competition at home and abroad.
- The technology structure should be advanced and suitable. The percentage of high-tech industry in total GDP will increase from 13% to 25%. The traditional industrial technology and innovation ability should be significantly improved.
- The distribution of regional structure should be reasonable and the comparative advantage should be obvious. With the policy of "improve the development of southern region, promote the rise of central region and use the advantage of backwardness of northern region," S Province will have the regional comparative advantages and form a reasonable divisional structure in the southern, central, and northern regions.
- The labor structure should be promoted. We should emphasize on the transformation of labor force of the primary industry, improve the employment level of the tertiary industry, promote the number of qualified management employees, establish

competition and talent exchange mechanism, improve total social security system, make the employment level stable, and ensure acceptable registered unemployment rates.

11.3.2 Choice of Upgrading Industrial Structure of S Province

S Province can be divided into three regions: southern, central, and northern regions. The southern region includes Suzhou, Wuxi, Changzhou, Nanjing, and Zhenjiang; the central region includes Yangzhou, Taizhou, and Nantong; the northern region includes Xuzhou, Lianyungang, Yancheng, Huaiyin, and Suqian. The economic gap between southern and northern regions is quite large, and the imbalance problem is serious. Briefly, the condition of development of the three regions are uneven and they are in different stages of industrialization. The advantages of resource and industry in different regions are different, and the condition, target, and key points of upgrading industrial structure are also different. We should differ in our views on the choice of upgrading industrial structure in different areas.

11.3.2.1 Approaches to Industrial Structure Upgrading in the Southern Region The southern region should accelerate the application of high-tech and relevant advanced technology (microelectronics and numerical-control technology are key areas). We can totally change the production methods and the approach of traditional industries by introducing new concepts, imitating and transforming. The distribution should be suitable, the development of high-tech industry should accelerate and the high technology should infiltrate into township enterprises. We can optimize the combination of the basic industry groups and promote the construction of heavy chemical industry bases in Nanjing, Zhenjiang, and Changzhou. We should accelerate the steps of information technology, package some potential products using information technology and increase the output proportion of high-tech industries of S Province.

The emphasis on the transformation of southern region is as follows: First, we should do the complete processing of raw materials which can increase the rate of finished products and extend the industrial chain in metallurgy, oil, chemical, and building materials.

At the same time, we can transform technology devices, save energy, and enhance the quality and technology of products which can promote competitiveness of products. Second, we must achieve electromechanical integration of equipment industry. While processing using microelectronics technology, we can mainly solve the patching problem of CNC systems in order to supply efficient, high quality, low power consumption devices. Third, the general processing industry should develop new products which can meet the requirements of the market to increase the market share. Fourth, the development of township should use suitable and high technology and flexible mechanisms to develop modern industries. Fifth, we should improve the scientific and technical application in agriculture. The industries that need transformation are machine industry, electronic industry, petrochemical industry, textiles and silk industry, light industry, metallurgical industry, and building materials industry as well as high precision electromechanical integration devices and high added value new materials.

According to the analysis of key points of upgrading industrial structure and the result of choosing leading industries of S Province, the southern region should mainly develop manufacturing industries on the basis of transforming traditional industries by high technology. Take the four state-level high technology development zones of Nanjing, Wuxi, Changzhou, and Suzhou and the Suzhou industrial park as carriers, implement the torch plan vigorously along the river, develop positively the electronic information technology, bioengineering, medicine, the new material industry, electromechanical integration which have comparative advantages, then form the Shanghai–Nanjing, high and new technology industrial belt; further, construct Shanghai line torch belt (Suzhou, Wuxi, and Changzhou) and Nanjing–Nantong torch belt (along the river) in order to promote high and new technology infiltration into the township enterprises; realize the upgrading of industry and transformation of traditional industries. During this process, the choice of industries, leading industries, and industrial distribution to the different cities in southern region are many (Table 11.12).

While using the high and new technology to transform traditional industry, we should speed up the information process. The key point is packaging potential products with the help information technology,

Table 11.12 Industrial Advantages and Basic Requirement of Each City in the Source Area

CITIES	BASIC REQUIREMENT	COMPETITIVE INDUSTRIES AND DEVELOPMENT MEASURES
Suzhou	Equipment manufacturing base for global information technology	Along the Chang river, heavy industries, processing and manufacturing center of the town, supporting products
Wuxi	An important modern manufacturing base in the Yangtze River Delta	High-grade textiles, metallurgical characteristics, household appliances, and so on
Changzhou	An important modern manufacturing base in the Yangtze River Delta	Agricultural machinery, power transmission equipment, vehicles and parts manufacturing, new textiles and clothing, and so on, electronic information equipment, new materials, biopharmaceuticals, fine chemicals
Zhenjiang	An important modern manufacturing base in the Yangtze River Delta	Basic raw materials, ship manufacturing (chemicals, paper, building materials, aluminum)
Nanjing	A research center, second only to Shanghai, for advanced manufacturing in the Yangtze River Delta	Petrochemical industry, vehicle manufacturers; building a world-class enterprise; increase high-tech enterprises in the industry share

raising the technique content, improving value addition of products; strengthening the transformation of leading industry with information technology, enhancing the production technology, management level and the labor productivity; realizing information technology in tertiary industry, promoting the starting and development of knowledge industry, and preparing conditions for spreading of industrial knowledge-base. We should positively use information resources, pay attention to the construction of economic, scientific, and technological aspects, construction of social nonprofit organization information database to realize the sharing of information. We should also pay attention to the information consulting industry, the information acquisition and processing industry and the information construction, information service industry as well as the information market construction. We can establish large-scale databases and convenient information network centers which can increase access of information to enterprises, individuals, families, public organizations; establish many large-scale information consultant service companies and continuously increase the proportion of information industries in the GDP to upgrade industrial structure.

11.3.2.2 Approaches to Industrial Structure Upgrading in the Central Region The northern region should speed up agriculture industrialization, other regions must accept the transfer of southern industries and cultivate leading industries. According to industrial development strategy of "four along," we should strengthen the regional centralization and the build-up effect, implement urbanization strategy, coordinate the proportion of three industry structure and the proportional relationship of high and new technology industry, the capital-intensive industry, and the labor-intensive industry.

The unbalanced development of economy divides the central region into the north zone and the south zone. Therefore, the upgrading of industrial structure in the central region should be carried out properly. The central region is rich in agriculture, aquatic and sea products, the management level of agricultural production is high, being the main production area of grain, cotton, kapok, oil, meat, and aquatic products. In Gao You, Baoying, Xinghua, Jiang Yan, Hai'an, Rugao, and Huangqiao, we must speed up the process of industrialization of agriculture. With the aid of the water resources and ecological advantages, we can develop characteristic agriculture, green agriculture, three-dimensional agriculture, tourism and sightseeing agriculture, which will convert the region into a support base to Shanghai and the southern region. We must develop extensive processing of agriculture and subsidiary products to increase the value addition, establish a batch of township enterprises or accept the transformation of southern township enterprises transfering the labor force remaining with the labor services for export and promote upgrading the industrial structure. Xinghua is a leading market for grain production; Qidong, Tongzhou, Rudong are leading markets for cotton and kapok; Hai'an, Rugao, Taixing are the meat production bases; Xinghua, Gaoyou, and Baoyingare are the aquatic products bases. In addition, spearmint oil, flowers, and trees crops also enjoy high reputation in China. Therefore, in the interest of the different regions, we should give proper guidance which can advance the process of industrialization of agriculture.

To some southern cities (counties) which have good industry foundation, we must reconfirm funds, resources, technology, brand and marketing networks, rejuvenate the industry by putting them into domestic and foreign competition. To those industries which have comparative advantage, good market prospects, quick growth, and strong connection effects, we must set up preferential policies, support the leading industry and create radically new growth. We can also take it as the main item which will lead the development of many necessary enterprises and related products. The controlling factor of supporting leading industries should be the leading products, which should have technical superiority and market superiority in a large scale. Of all industries, the top five of added value industries are textiles, electrical appliance machinery and manufacture, chemical fiber manufacture, standard machine manufacture, chemical raw materials and the chemical product manufacture which are still labor intensive. To regulate the storage quantity structure in the traditional industry in the central region under market mechanism, we must abandon the behavior of refraining from doing things. Those state-owned enterprises having no core competitiveness should quit and not be helped out. The existing pillar industries like automobile, motorcycle, air conditioning, petroleum, and chemical industries are mainly traditional industries; the proportion of high and new technology industries is low. But the superiority of these traditional industries is obvious, and the supporting effect is strong. Through technology acquisition, imitation, and innovation, the central region may choose to develop those industries which have high increment, require high intelligence, high actuation, high strategy, high risk and high effectiveness to regulate the economic structure and realize summary expansion and structural optimization.

To analyze the condition of industrial development of the central region and its characteristics, the central region must strengthen the upgrading with the regional industry axis as the center. The industrial belts of the central region are basically in the scale expansion and the nodal growth stage; the axis belt is not strong. But the tendency for the industry and population to gravitate toward the axis is quite high, especially to all grades of cities; the belt-shaped pattern of industry distribution is obvious. Depending on the foundation, resource superiority, direction of economic and spatial organization of various industries, the central region has formed four main axis belts: east–west basic industry belt along the river, the north–south industry belt along the river to have a coastal economic belt, the Beijing–Hangzhou canal economic belt along the highway of Nanjing–Nantong, and the economic belt of the Beijing–Shanghai highway. These are called the "four along" industrial belts.

1. *Development along the river.* This makes full use of the geographical superiority along the river and sea, and by using the two markets, domestic and foreign, two resources, taking Nantong port, Gaogang, and Yangzhou ports as backups, advancing east and west together, and spreading to the north. We should lay out foundation materials industries, iron and steel, petrochemicals, building materials, and large-scale process manufacturing industries that have big freight volume and large water consumption. Form the Bin Jiang basic industry belt and construct the basic industry system that adapts to the Yangtze Delta economic zone assembly processing industrial structure. At present, besides shipping, energy, and heavy chemical industries of Yangzhou, fine chemicals of Taizhou, chemical industry, building materials, metallurgy, and the relative industrial superiority of Nantong, we must take the alkenes (the aromatic hydrocarbon) as the main item, form new nationwide chemical base image that has high technique content, big economic scale, with petrochemicals for the leading industry and fine chemical industry for characteristic industry.
2. *Development along the sea.* Among the middle resources, the status of marine resource is prominent. The central region must consider the realistic possibility of development, increase the developmental dynamics and speed on comprehensive plan in order to raise the developed shoals. We must take the opportunity to construct deep-water harbor Lv 4 port, Yangkou port, lead by developing electric power, petrochemicals, iron and steel, sea industry, and form the coastal industrial belt. We must unearth the local tourist resources, improve the infrastructure for tourism, develop the tourist route, improve the service grade, play the feature tourism sign and transform superiority of resources to industrial superiority as soon as possible. We must vigorously develop the physical resources of the sea while developing traditional industries such as oceanographic catching and aquaculture; we should

depend upon high and new technology development industries such as marine life and medicine, sea minerals, sea chemicals industry.

3. *Development along the road.* We should form "two horizontal two vertical" industrial spool thread along the road by Nanjing–Nantong, Nanjing–Qidong, Nanjing–Shanghai, Xin–chang railway. This industrial spool thread intercrosses with the developmental strategy along the river and key city development, specially with the Jiangyin bridge, Runyang bridge, and Sutong bridge which form the highway and bridge economy. There are two aspects. First, establishing the industrial belt along the road and the bridge can aggregate capital, technology and information, display build-up effect, speed up modernization process of rural areas along the highway and bridges. Then display the diffusion effect that leads the development of peripheral localities. Second, through highway and bridges we can establish many kinds of convenient channels for strengthening contact with Shanghai and the southern region, minimize the transportation cost, create conditions for enhancement of the output of the central region and expansion of the domestic and foreign markets. At the same time, convenient transportation reduces the spatial difference, speeds up production fluidity, creating more advantageous conditions for accumulation of all elements and transportation to Shanghai and the southern industries.

The economic globalization has made possible resources integration, technical integration, information integration, the market integration, and the industrial integration globally, and so on; the economic globalization intensified the interdependence of the economy of various countries, which has provided external conditions for upgrading the industrial structure of the central region. The central region may face the world market, participate in the international division and the international exchange positively, expand opening to the outside world, implement the strategy of "introduce in" and "walk out," use own superiority to attract more funds and technology, promote export of characteristic agricultural products (such as gingko, lotus root, flowers and trees, fish and shrimp, goose duck, intensive processing of paddy rice products, etc.), export of labor intensive products and labor services, thus realizing upgrading of the industrial structure.

11.3.2.3 Approaches to Industrial Structure Upgrading in the Northern Region The northern region should take to promoting the level of industrial structure of agriculture to breakthrough point, rationally distribute agriculture productive forces structure and positively advance the agricultural industrial production. With the aid of Shanghai, southern region, and overseas technology, the northern region should display the duplication effect and the crossover effect. It should also transform traditional industries such as food, machinery, chemical industry, building materials by high technology, speed up the development of ventures in agro-industry, develop labor-intensive industry, speed up the rural labor force transformation vigorously.

Agriculture is the basic industry of the northern region. Agricultural resources is a huge advantage of the northern region, which is not only the economic main body of northern region, but also the main agriculture region of the entire province. One way of upgrading the industrial structure of the northern region is to advance agriculture production on industrial scale. The northern region should first cultivate the leading enterprises and the enterprise groups positively, take leading enterprises as the central ones, centralize financial resources and physical resources to cultivate leading enterprises which have large-scale and extensive spread. At the same time, the northern region should encourage nonpublic ownership system enterprises to enter into the domain of agricultural development, promote the industrial production to a higher level. The northern region should also further cultivate industrial crop base of Xu Huai, Lixia River, along the coast, as well as establish fruit farms, aquaculture, flowers and plants, vegetables, dried mushrooms, Chinese medicine industry, and raising livestock base by using natural resources in order to improve and encourage agriculture products.

The northern region has mineral resources, occupies half of agricultural farming area and the marine resources of S Province; the potential for developing resources processing industry and the sea industry is huge, but its technical and industrial exploitation level is not high, so it should introduce high and new technology technology similar to that applied in Shanghai and in southern region as well as overseas to expand production capacity and realize the industrialization of the northern region and transform the traditional industries of the northern region like food, machinery, chemical industry, and building materials. The northern region may also introduce scientific achievements as in Shanghai, southern region, and overseas, thus reducing the development gap with the industries of Shanghai, southern region as well as that of overseas industries.

The northern region should maintain the significant transformation of agricultural development strategy after joining WTO, and the opportunity to join the international division system. Then it should accelerate developing agricultural processing industry to promote industrial competitiveness.

1. *Understand the significant strategic importance of developing ventures in agro-industry.* Raise the combined agricultural earnings, the importance of market competitiveness regarding the expanding agricultural product market demand, lead the agricultural restructuring to increase the farmers income. The northern region should take the market as a guide, display the basic function of market mechanism of agro-industry development ventures, the basis of developmental strategies of sentiment, actual formulation of agricultural products processing, industrial structure, and environmental protection. Take the scientific innovation as the power, take the township enterprises as junction points, take the leading enterprises as the benchmarking display the technology and the resources advantages to realize the ventures in agro-industry of the northern region by preliminary work in intensive processing, from rough machining to precision work, and then advance ventures in agro-industry development will reach new levels.
2. *Create free policy environment for ventures in agro-industry development.* Increase the investment efforts, support the high quality processing and the transformation of agricultural products positively, and reinforce the government investment by investing in high technique content projects. Expand the equity system reform, realize the subject of investment multiplication, make preferential policies, create good investment environment, and increase the dynamics of public funding investment.

Enlarge the financial organ support dynamics depending on the ventures in the direction of development of agro-industry reasonable adjustment loan structure, increase the loan total quantity suitably, and positively expand the financing channel of large-scale agricultural products processing enterprises. Formulate the corresponding tax-preference policy. Unify the development ventures in agro-industry and environmental protection, reduce the processing of agricultural products as far as possible to avoid environmental pollution, and realize sustainable development. The foreign merchant investment environment should be improved, the kinds of preferential policy should be adopted so that it attracts more of foreign investment for expanding attractive investments domains, guide and encourage the international capital to adopt different ways to go to the ventures in agro-industry.

3. *Using the opportunity of entering WTO to build new competitive advantage.* The R&D ability is still the weak link of the ventures of S Province in agro-industry's professional competitive power. In 2001, the new output only accounted for 4.9% of total output, 7.9% lower than the complete industry proportion, the superiority of process technology has not been manifested. The northern region should unceasingly enhance the technology, craft, and management level of agricultural products processing, encourage the enterprises to develop new superiority in research and development, brand building, organizational structure to shape stable core competitiveness. On the one hand, the northern region should continuously introduce the overseas vanguard technology, craft, equipment, and management techniques. On the other hand, it must urge large and middle scale village industry enterprises which have strong economic power and technology base to establish technological innovation systems as soon as possible, form the innovation mechanisms of enterprise-independent development as well as develop technological applications with the help of scientific research institutions and universities, enlarge the application of high technology and new technology, speed up technological transformation step of enterprises. It should also study the demand of international market, take measures

to support the agricultural products processing enterprises in facing the international market growth production, encourage and support those strong enterprises to encompass international market, pay attention to develop characteristic international brands within S Province, expand self-management export.

4. *Realize the development of ventures in agro-industry.* The northern region can develop the industrialization of agriculture with "company + peasant," form, use the branching effect of leading enterprises to connect agricultural production with processing, maximize the common benefit, thus strengthening the connection between agriculture and the ventures in agro-industry. The unification of agricultural production and processing enables the formation of the management mechanism by which agriculture and the ventures in agro-industry can mutually promote each other, break the department division and monopoly, and make an industrial system with agriculture as a competitive power. The northern region should cultivate agro-industry as a leading enterprise, take the development of agro-industry as an important undertaking of rural enterprises, carry on intensive processing of agricultural products with regard to agriculture as leading industry with superiority of products, form a system whereby the production, circulation, and processing are integrated, and speed up the formation of agriculture as a superior industry.

Positively develop labor-intensive industries, speed up the transfer of rural labor force. The proportion of population involved in agriculture in the northern region is quite large, which has restricted S Province from upgrading its industrial structure, specially the improvement in the structure of labor force. We must develop the labor-intensive industry to increase the absorbtion of labor force in the economic system and create convenient conditions for labor force flow. First, further drive the promotion of city industrial structure in order to create more employment opportunities and create suitable conditions for massive employment of labor; second, encourage uniform purchasing of farm produce by the state according to fixed quotes to nonagricultural production. Only in this way can industrial structure of agriculture realize a higher level of formal management and attain conditions for improvement. Southern cities like Nanjing, Suzhou, Wuxi, and Changzhou are highly open and have strong industry foundation; they should enlist rural labor force from the northern region on their own initiative.

11.4 Countermeasure of Realizing the Upgrading of Industrial Structure in S Province

11.4.1 Establishing the Scientific Development View, Further Enhancing the Understanding, Creating Good Environment for Upgrading Industrial Structure

The upgrading of industrial structure is a complex process. It not only involves huge resources (people, capital, material), but also the structure promotion adjustment for the benefit of department, region, enterprise as well as personnel. Therefore, the leaders and the enterprise managers should come together and establish a scientific development viewpoint for the sake of long-term development and that of the economic development, and correctly understand the contribution of industrial structure upgrading to the economic and social development.

Technological innovation is an important way to upgrade the industrial structure. When the government organizes technological innovation activities, it should further transform the function to the level of macroeconomic control, creating the condition and the environment, formulating law and regulations, providing policy instructions and services, promoting cooperation in product study routines. In order that enterprise technological innovation information is unhampered, the government must increase the dynamics of infrastructural investment of industry technological innovation information and provide effective sources of information for enterprise technological innovation activities. It should also strengthen the innovative ideology, propagate the entrepreneurial spirit, unify the understanding of the importance of technological innovation.

The government should take early measures to strengthen the function of laws and regulations, to prevent disruption of the market by foreign capital control or abuse its superiority, of system construction. It should maintain fair competition and standard competition behavior.

11.4.2 Guaranteeing Steady Development and Advancing the Upgrading of Regional Agriculture Industrial Structure

S Province is big-agricultural province. The development of rural economy is one of its economic development missions. Therefore, we must insist on strengthening the leading position of agriculture in the first place. S Province has a large population and little land, the effective supply of insufficient agricultural products will exist for a certain time; in addition, the foundations of agriculture is frail, factors that negatively affect the steady progress in agriculture have not been alleviated. The highly effective and stable development of agriculture will depend on the work dynamics: to increase investment in agricultural materials, technical support, and capital; to carry out policy measures of protecting and supporting agriculture, to improve condition of agricultural production, to strengthen agriculture infrastructure, to depend upon the advancement in technology development for high production, high quality, and highly effective agriculture. When formulating plans for upgrading industrial structure, we must arrange the agriculture in a proper manner and support agriculture industry first.

In agriculture, when developing the diversified management and upgrading industrial structure, we cannot relax food production. On grounds of potential losses on aquatic products, forest fruits, tea foundations, we should further stress the layout plan, seedling stock breeding, feed and the guard system construction, base construction, and give support to the credit fund to enable diversified management. The grain and cotton are the key agricultural products of S Province; therefore, in the production of grain, cotton, and kapok, we should further implement "let industry support agriculture," and "to construct agriculture by the labor" system and construction measures, and so on. Productions of cotton and crops play important roles in the northern S Province. Next, the local government should encourage farmers to plant crops and cotton and to get a sustainable development.

The development of S Province area is not balanced; we should act according to the conditions of southern, central, and the northern regions, formulate target-oriented strategy of upgrading the regional agriculture industrial structure development and advance the upgrading of agriculture industrial structure step by step.

11.4.3 Multi-Pronged and Uneven Advancement of the Upgrading of Secondary Industrial Structure

11.4.3.1 Strengthening the Macroscopic Management, Concentrating Efforts, Limiting Objectives, Prominent Key Work out a medium and long-term industrial structure upgrading development project from the macroscopic angle, take the industrial structure upgrading of achieving "11th ten year" as the key point, draws up sets of measures in finance, tax revenue, and finance. Choose some significant key technologies of industrial development, centralize the manpower, physical resources and financial resources to carry on in order to raise the technical level of the industry rapidly. In view of the different characteristic of the industry, we should take different policy measures. The current key point is to enhance the technological innovation ability of the equipment industry as well as to cultivate some large enterprise groups which have the technological innovation ability and international competitiveness.

11.4.3.2 Advancing the Dynamics of Enterprise By advancing the dynamics of enterprise, develop the technology and technological transformations, increase the investment proportion for the technological transformations and the high and new technology industry, speed up the upgrading of industrial technology system through the following.

1. Through policy establishment of key industry fast depreciation system and promotion of enterprise modernization, such as carrying on guaranteed or reduced interest to some shareholding system or important professional equipment investment loan, promoting the updating of the equipment of the enterprises, formulating professional technical standards, eliminating obsoleteness in craft and technical equipment.
2. Formulate policies that are beneficial to increased investment in the enterprise technology development innovation, and realize transformation from technology development by the government resulting in leading enterprises. The government should subsidize technology development of some essential domains, exempt them from income tax, include the production cost in intangible asset's investment; request that the

government provide financial support to some development facility for research and have certain projects unify with the enterprise.

3. Take the path which enables technology introduction, assimilitating absorption, and independent innovation; increase the technical source for multichannel traditional industry transformation; enhance the dynamics of the technology acquisition, introduce domestic and foreign vanguard technologies positively, reduce the disparity in industry technical and the domestic and foreign vanguard technology; strengthen the macroscopic management of technology acquisition; prevent redundant, blind, and low level introduction.
4. Strengthen coordination ability between various departments. The government must give support in funding and technology and act to organize key technical aassimilation, improving the domestic product of some key equipment manufacturing.
5. Establish agency for technology introduction and evaluation; help enterprises enhance their technique and ability; establish a network of technically skilled people, help small and medium-sized enterprises to carry on technical diagnosis and enhance technical service ability.
6. Establish technological innovation mechanism for inner power and outside pressure; give importance to technological innovation; make the enterprise attract risk investment, and benefit the main body of technological innovation. Realize the transformation of S Province enterprises from traditional production configuration enterprises to innovation enterprises.

11.4.3.3 Cultivating Industry Colony and Strengthening the Regional Industry Innovation Ability The industrial group is the spatial gathering of industries which have close professional, economic, and technical relations. It is the organization cell of regional industrial structure rationalization and upgrading. Macroscopically, it displays regional industrial structure including the leading industry, support industry and the economic basis; microscopically, it displays as high division of labor and cooperation and specialized production between profession and department. The formation of the industrial group is advantageous to form the "local network" which has stable close relation, and the enterprise that takes root in the local society culture with the localization characteristic, namely, "rooted in nature," enables the enterprise to innovate unceasingly and have self-development ability. The innovation, competition, studies, and references inside the industrial group will certainly enable technological innovation and technology advancement, thus providing impetus for industrial structure upgrading.

11.4.4 Positively Developing the Emergent Industry and Realizing the Upgrading of Tertiary Industry Structure

11.4.4.1 Adjusting Investment Trends, Carrying on Investment Guidance according to the Collective Services, Half-Collective Services, Personal Services The sensitivity coefficient of tertiary industries is not high, athough its influence coefficient is quite high, which indicates that the tertiary industry development is next only to national economy development and its structure upgrading effect was inferior to the effect that promote the national economy development and structure upgrading. Therefore, the tertiary industry of S Province is suitable for adopting the strategy of development initiative and structure promotion, and it is the best way to increase the investment in tertiary industry. Moreover, as it has been implied that investment in the tertiary industry is smaller than in developed countries, S Province must adjust the investment trends. Before giving investment guidance, the basis of solving this question is the three different kinds of service industries.

Besides providing collective services, nonprofit collective services and the control monopolization service industry (for example, postal service industry, telecommunications industry, financial industry), other professions, especially the professions which provide personal services, should be open to public and the main body for investments abroad. We should encourage foreign capital to enter tertiary industries, specially those key industries according to "who invests, who possesses, who profits" principle. The government must enhance investment in collective services profession, the nonprofit collective services profession and service industries with low self-developing power such as farming, forestry, herd fishing service industry, warehousing industry, education, and healthcare business.

11.4.4.2 Implementing Innovation of Scientific Services, Positively Introducing New Technology of Service, Promoting Modernization of Tertiary Industry and Increasing the Provision of New Service Products We may know from the low sensitivity coefficient of the tertiary industry of the S Province that the demand for service industry is not fierce because of low income levels, low socialization of production and low specialized degrees. As a result, we should introduce positively new technology of service, advance upgrading of tertiary industry, stimulate new service demand by increasing new service products supply.

The development of modern technology, especially the high and new technology, has brought revolutionary influence to the tertiary industry; for example, the electronic information technology, the Internet technology, and the integration of machinery technology have huge influence on the tertiary industry information services; online shopping, distance learning, and long-distance medical services also provide the tertiary industry more development opportunities. Therefore, the tertiary industry of S Province needs to introduce vanguard technology, enhance the modernization level of the tertiary industry, increase supplies of new service products, promote tertiary industry structure upgrading by increasing new products and stimulating new demand.

11.4.4.3 Loosening and Enhancing Quality Price Difference of Service Products The Chinese price change of tertiary industry products has lower influence on the primary industry, secondary industry, and agriculture product price than in the developed countries; it has also lower influence on other industrial price stability than the secondary industry. Therefore, we should do away with the price control of the service products, and adapt the differential level, the differential local service consumer demand through a suitable expansion of service product's quality price and the regional price difference and increased output of the tertiary industry through price enhancement of the service product.

Gradually reduce the range of price control of service industry by (1) completely opening up the price control of private service industry to let it adjust by the market forces and (2) doing away with the price control of the half-collective industry, especially the natural monopoly industries such as post telecommunication and air transportation.

Properly enhance the service product quality price by (1) making different rank price of service product according to "high quality favorable price, poor quality poor price" principle and (2) implementing control of service "the quality price difference rate," stipulate reasonable "quality price difference rate" to prevent enhancing excessively the service product quality price difference.

11.4.4.4 Vigorously Cultivating Emergent Industries and Adjusting Internal Structure of the Tertiary Industry During the period of switching from industrial economy to knowledge economy, many new industries will arise to adapt to knowledge innovation, dissemination, and application. Inside the system of knowledge innovation, the education industry will arise; inside the system of knowledge dissemination, the network industry, dissemination industry will arise; inside the system of knowledge application, the consultation industry participating in the management service industry will get high value addition. Moreover, to adapt to the new change brought by knowledge economy, healthcare profession of high-tech content will also get much developing space. We must utilize different economic levers to the emerging industries including tax revenue, credit, and finance in order to guarantee fast development of these professions. We can increase the demand for service products by supplying more, so as to upgrade the industrial inner structure and optimize the structure of the three industries.

11.4.5 Advancing the Step to Develop High and New Technology Industries

11.4.5.1 Implementing Major Scientific and Technological Special Projects to Enhance Scientific and Technological Innovation The special project of significant technical should consider scientific innovation as the core, constructing the international manufacturing industry base as the goal, with regard to the construction equipment manufacture (automobile, ships, engine bed, complete set of equipment is key point, development integration of machinery equipment), the chemical industry (petroleum chemical industry, fine chemicals, composition material chemical industry chain), the metallurgy, the physical distribution for big industry colonies, as well as the electronic information, the machinery, the metallurgy, the petroleum chemical industry, the spinning and weaving—five big industrial bases, closely hold the medicine industry, the automobile, the environmental protection, the food. These four key industries to make further development, realize the upgrading of industries and the industrial structure diligently melting, make science and technology and the production of S Province and force "the nature" to leap high.

The integration of core science and application of technology will be good for electronic information, chemical engineering, bioengineering, new medicine, and new material industry which have comparative superiority, for core product group including new microelectron product, modern communication product, computer and the network product, digitized visual and hearing product, nonmaterial optical communication material, rare earth compound materials, electronic information materials, and new medicines. Under the guidance of integrated thought, we construct community superiority and partial superiority of high technology and new technology industry.

11.4.5.2 Establishing High-Tech Industry Development Area and Cultivating High-Tech Industry Belt along the River The eight cities which are along the river, especially Nanjing, Suzhou, Wuxi, and Changzhou, are the main locations of social, economic, and technological development. The industry class of this region is complete, the secondary production superiority is obvious, which has over 80% of total output and research ability of mechanical and electrical industry, spinning and weaving, silk industry, chemical industry, medicine industry, metallurgy, and building materials. It is also the production base of textiles and silk, which has concentrated the entire province majority of vanguard technology equipment and the high-tech product. According to their foundation and high and new technology potential, the eight cities can be divided into two development levels: The first level includes Suzhou, Wuxi, Nanjing, and Changzhou; the second level includes Nantong, Yangzhou, Taizhou, and Zhenjiang. At present, we must vigorously develop the first level, speed up developing the second level to emphasize on key points and give dual attention to the overall situation. Then we choose certain industries having high industrial correlation, broad market prospects, high-technology content, tremendous project influence for the entire economic development of the province according to regional situations in technical level, economic efficiency, development potential. Take them as the key points of high technology and new technology industrial production, create well-known branded goods which have the proprietary intellectual property rights in order to complete the leading economic development of the area, to lead the northern regional promotion in promoting the Yangtze valley opening to development, to have "along the river" comparisons of the market competition ability of high technology and new technology industries inside and outside the powerful nation.

11.4.5.3 Developing Product Groups of High Technology The so-called high new technology product group refers to the development of a series of relatively high new technology products which will form an accumulative effect. This kind of development does not necessarily carry on in one high-tech enterprise, it may carry on in many high-tech enterprises, and these high-tech enterprises will help each other. When cultivating high new technology product group, the standpoint lies in transforming the regional and professional superiority to product superiority. The key points are electronic information, integration of machinery, new material, bioengineering, and the new medicine industries. Moreover, S Province has the foundation for developing new technology product groups in numerous aspects, has developed a lot in high-tech industry torch belt along the river, has superiority in the market of high new technology products, which will have good prospects of development.

11.4.5.4 Developing a System of Risk Financing and Cultivating Small and Medium High-Tech Enterprises The high-tech companies have high innovation, high utility, high risk and benefit, so the banks often reject giving loans. This needs establishing a financing system which can undertake the risk. The venture capital which requires high repayment and undertakes high risk is precisely the financing support system. It helps the high-tech company's product to move from the laboratory toward the market.

11.4.6 Adjusting Strategy of Introducing Foreign Capital, Enlarging Dynamics of Bringing in Foreign Capital and Speeding Up the Industrial Upgrading through the Introduction of Foreign Capital

The industrial development is not balanced in S Province; the southern, central, the northern regions are at different industrialization development phases. So we should have different actions when introducing funds and prompt economical development both the southern and northern regions in S Province. The southern region should focus on enhancing transformation ability of industrial upgrading, promote upgrading industrial structure with foreign capital. Put in prominent position the pursuit of promoting technology advancement, industrial evolution, the mechanism transformation; implement "the industrial structure upgrading guidance" strategy of bringing in foreign capital. The northern region is left behind in economy; the industrial level is low and technical level is also backward, so it must encourage bringing in foreign capital, speed up the implementation of "capital formation guidance" strategy to fix its economic role in the Yangtze Delta region, pay great attention to grasp the market pulse, display comparative superiority, speed up the infrastructural facilities, take the opportunity for transforming traditional industry into developed ones, strengthen industrial connection with Shanghai, the southern region, Hangzhou bay region and be clear about its own industrial localization.

1. Strengthen industrial guidance foreign investment, promote industrial structure optimization. Be definite about the industrial guidance, reasonable plan, correct guidelines, and expand the domain of using foreign funds in the tertiary industry according to the trend of industrial evolution and development key. Implement special and preferential treatment to high and new technology industry investment, display the hatching and aviation effect of high and new technology investment, enhance the industry development and innovation ability to promote international competitiveness.
2. Further expand the way of introducing foreign capital and open more ways to use foreign capital. Permit foreign capital to take part in reorganization or enterprise merger and acquisition through property rights transaction of state-owned business and non-state-owned business assets; while continuing to pay special attention, borrow foreign loan, and try to develop securitization financing. At the same time, we must improve soft environment of investment, establish foreign merchant investment management mechanism and the service system that conforms to the international convention and the domestic regulations.

11.4.7 Completing the Mechanism and Policy of the Leading Industry Development, Establishing Ways of Realizing the Leading Industry

Establish the leading industry development mechanism through market regulation and the government. On the one hand, the government speeds up the upgrading of leading industry by supporting them; on the other hand, the industry adjusts spontaneously with other industries through market mechanism function, which enables the leading industry to have intrinsic development request and good development environment. It is advantageous for the healthy development of leading industry.

Implement the industrial policy which encourages the leading industry development. Every country has the industrial policy to encourage their own leading industries. S Province may profit from the following: first, the financial policy, including the investment grants, accelerates to amortize, reduce tax revenue; second, the financial policy, namely, stipulation comparison preferential benefit's commercial present interest rate and the big scale's loan specified amount, which can support the social capital to go to the leading industry; third, the foreign trade policy conforms to the WTO regulations to support the import–export trade of the leading product.

Establish ways and channels to realize leading industries. There are two ways: first, the market adjusts spontaneously. Generally, the competition and the supply–demand relation can promote the competitive ability of industries. Upgrading of industrial structure may also be realized by the market supply and demand and the price mechanism. At the same time, the market regulation has the highest efficiency, the lowest cost, and the best authority also. But the "market malfunction" exists. The market mechanism will sometimes be unable to play its role. Second, government rules and regulations drive the industry economy not only because of natural reasons such as natural monopoly, information unsymmetrical, exterior nature, public goods, which make the market mechanism inextricable, but also the reason of social justice and ideology; for example, to weaken an excessively formidable economy and political rights of the monopoly enterprise by limiting monopoly, maintain the employment opportunity fairly through work, employment rules and regulations. Both mechanisms should be rationally divided, organically coordinated; the market mechanism has foundational function, and the government coordinated mechanism focuses on maintaining the operation of market mechanism and overcoming market failure. Only coordinating the two mechanisms can guarantee unimpeded opportunity for realizing leading industries.

12

Study on Ways of Industrial Structure Advancement during the 11th Five-Year Plan in China

12.1 Main Problems Existing in Optimization of the Industrial Structure

Since reform and opening up, our economy has stepped into quick growing period and the GDP has increased from 362.44 billion RMB in 1978 to 11669.4 billion RMB in 2003. Our present industrial structure and employment structure are shown in Table 12.1. We can see that the three industrial structures have changed significantly. The GDP of the primary industry has declined rapidly, that of the secondary industry has improved and is stable, and that of the tertiary industry has increased slowly. This situation is nearly the same as that of the employment structure. All these show that our industrial structure is progressing in quantity toward an optimized way. However, the optimization of industrial structure appears not only in quantity but also in quality, especially in the post-industrialization society, where the appearance of knowledge economy, the change of economic growth mode, and the optimization of the industrial structure become imperative. It will influence the whole situation of development in the national economy. At present, the main problems of optimization of industrial structure will be discussed.

12.1.1 Unreasonable Industrial Structure

First, the proportion of the primary industrial output value in the national economy has declined too fast. Second, the proportion of the

Table 12.1 Industrial Structure and Employment Structure since 1978 in China (%)

	PRIMARY INDUSTRY	SECONDARY INDUSTRY	TERTIARY INDUSTRY
GDP STRUCTURE			
1978	28.1	48.2	23.7
1985	28.4	43.1	28.5
1995	20.5	48.8	30.7
2000	16.4	50.2	33.4
2001	15.8	50.1	34.1
2002	15.4	51.1	33.5
2003	14.8	52.9	32.3
EMPLOYMENT STRUCTURE			
1978	70.5	17.3	12.2
1985	62.4	20.8	16.8
1995	52.2	23.0	24.8
2000	50.0	22.5	27.5
2001	50.0	22.3	27.7
2002	50.0	21.4	28.6

Source: Statistical Yearbook of China in 2003.

secondary industry output value is too large; and the internal structure is unreasonable; improvement of the basic industry and the fabrication industry is uncoordinated. Third, the proportion of income and employment in the tertiary industry is low, especially in transportation industry, information industry, finance and insurance industry, and service industry, which are important components for good modern living. Our agriculture is in a low and basic position and the modernization level is low. The tertiary industry lacks full development having started relatively late as a new industry. So, the development level is still low.

We should not depart from the employment structure when we investigate the industrial structure adjustment. We should connect the important indexes of the two parts together. We define valuable metrics called average increment of industry to evaluate the contribution of an industry to the national economy. C_i stands for the contribution to the national economy by average increment of industry i. The formula is as follows:

$$C(i)=\frac{\text{GDP of industry } i}{\text{Employment in industry } i}, \quad C_i=\frac{C(i)}{\sum_{i=1}^{n} C(i)}, \quad i=1, 2, \ldots, n$$

Here $C(i)$ is the industrial comparative labor productivity which can be used easily to transversely compare the internal industrial productivities in a country. However, it is hard to see the difference from the transverse comparison between different countries owing to the difference in the total amount in the economy. So we standardized the industrial comparative productivity of each country or area to get C_i. This metric can first integrate the industry and fabrication structure to investigate the actual industry structure. Second, it can show the differences in productivities between developed countries and China; and it also shows the differences in relative relations corresponding to the three major industries. Besides, it can reflect the comparative status of each industry in the national economy. Based on the data of 2000, we calculate the contribution degrees of industrial average increment to the national economy of USA, Germany, Japan, South Korea, and China. The original data is shown in Tables 12.2 and 12.3.

From Table 12.3 we can see that the average contribution degree of the secondary industry to the national economy is the highest; on the contrary, in case of primary and tertiary industries it is very low, especially for the primary industry. The average contribution degree of the primary industry is only one-seventh of that in the secondary industry. However, the above proportion relation is four-seventh in USA, one-half in Germany, one-fourth in Japan, and two-seventh in

Table 12.2 Industrial and Employment Structure for China, USA, Germany, Japan, and South Korea in 2000 (%)

	PRIMARY INDUSTRY	SECONDARY INDUSTRY	TERTIARY INDUSTRY
GDP STRUCTURE			
China	16.4	50.2	33.4
USA	1.6	24.9	73.5
Germany	1.2	31.5	67.3
Japan	1.4	31.8	66.8
South Korea	4.7	42.4	52.9
EMPLOYMENT STRUCTURE			
China	50.0	22.5	27.5
USA	2.6	22.9	74.5
Germany	2.8	34.5	62.6
Japan	5.1	31.2	63.1
South Korea	10.9	28.0	61.1

Table 12.3 Degree of Contribution of Per Capita Industry to the National Economy for Related Countries in 2000

COUNTRY	C(1)	C(2)	C(3)	C_1	C_2	C_3
China	0.3280	2.2311	1.2145	0.09	0.59	0.32
USA	0.6154	1.0873	0.9866	0.23	0.40	0.37
Germany	0.4286	0.9130	1.0751	0.18	0.38	0.44
Japan	0.2745	1.0192	1.0586	0.12	0.43	0.45
South Korea	0.4312	1.5143	0.8658	0.15	0.54	0.31

South Korea. So we can conclude that agriculture is in an unreasonable status in our national economy structure and a serious situation exists in other industries. From a comparison between the secondary industry and the tertiary industry, we see that the average contribution degree of the tertiary industry is half that of the secondary industry; also the above proportion relation is 1 in USA; six-seventh in Germany; nearly 1 in Japan, and four-seventh in South Korea. So the tertiary industry is also weak in our country.

The contribution degree of the secondary industry to our national economy is in an absolutely preponderant status and it is 20% higher than the average level of developed countries. However, it cannot prove that our industry structure is strong. Actually although our industry size is considerable, the industries lack of relative competition abilities; resource-intensive and labor-intensive corporations account for the most in our country; the ability of industry structure for independent development is weak and production structure is unreasonable.

12.1.2 Uncoordinated Development of Industry and Lack of Competitive Ability

First, the development speed of the three major industries is uncoordinated. From the analysis of increment of the three industries since 1995, we find that the gap in increasing the development speed between the primary and the secondary industry is increasing, which is seven-eighth in 1996 and one-half in 2002; obviously, the development speed and the after effect of the primary industry is much smaller than that of the secondary industry (Figure 12.1).

Second, the investment of the three industries is uncoordinated. The total investment for the infrastructure in 2002 was 1725.13 billion

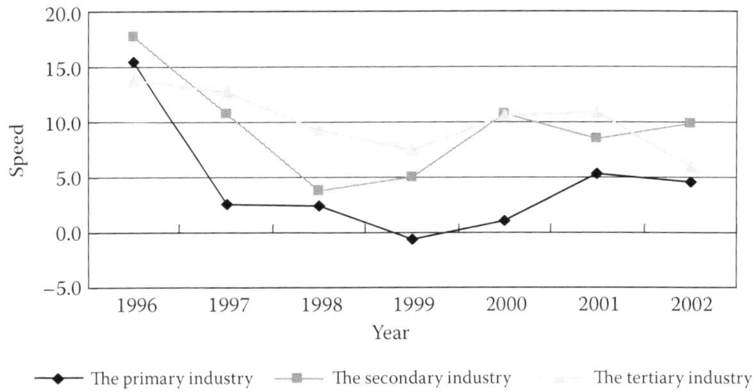

Figure 12.1 Development speeds of the three major industries in China (in terms of comparative price calculation).

RMB and it is only 101.84 billion RMB for the primary industry, which is about 5.9% of the total infrastructure investment. The total industrial investment for renewal and reconstruction is 658.42 billion RMB, and it is 5.12 billion RMB for the primary industry, which is about 0.8% in the total industrial investment. So, although the basis of the primary industry is weak, the investment and reconstruction degree for the primary industry is not prominent and this situation has hardly changed in a long time. Besides, the investment inclines excessively toward corporations with nonideal economic benefit indexes such as state-owned enterprises, and the problem of hard financing for small and medium-sized and private enterprises is not completely solved. Since 2000, the investment for state-owned and other fixed assets account for over 70% of all social investment and is increasing. But the increase in the proportion of collective and individual investment is lower than that in state-owned economy. In conclusion, the uncoordinated investment is one of reasons for uncoordinated development speed (Table 12.4).

12.1.3 General Low Efficiency of Industry Structure

Nowadays, economy increase depends not only on investment of capital, labor, and technology, but also on whether economic resources are optimized. Industry structure status and evolution decide the effect of allocation of resources to a large extent. Industrial structure can

Table 12.4 Investment Structure of Fixed Assets since 1990 in China (%)

YEAR	NATIONAL INVESTMENT VALUE OF FIXED ASSETS (10^8 yuan)	ECONOMY		
		STATE-OWNED AND OTHER	COLLECTIVE	PRIVATE
1990	4517.0	66.1	11.7	22.2
1995	20019.3	70.8	16.4	12.8
2000	32917.7	71.1	14.6	14.3
2001	37213.5	71.2	14.2	14.6
2002	43201.0	71.8	13.7	14.5
2003	55118.0	72.1	14.2	13.7

be seen essentially as a resource converter. But the operating efficiency of our resource converter is not high. Asymmetric production and demand are the direct results from inefficient industrial structures. Chinese traditional industries have large production capacities. However, the technology is backward, material consumption is too high and products cannot be well positioned to meet the growing needs of the people in our traditional industries. Traditional industries which are not updated and improved in time will easily result in premature recession and will affect the healthy development of the national economy. Our traditional products, especially from the labor-intensive industries and the low-skilled ones, are identical in grade and level; the production is relatively too large and will cause vicious competition both in domestic and international markets.

Inefficient industrial structure is also reflected in the structure of the input and output asymmetry. From 1990 to 2003, China's fixed assets grew nearly 12 times while GDP growth was less than 6.3 times. The effect of coefficient of investment in fixed assets declined from 58.5 RMB/100 RMB in 1995 to 22.3 RMB/100 RMB in 2000 and to 17.4 RMB/100 RMB in 2001 and 15.0 RMB/100 RMB in 2002. This index can approximately reflect the macroeconomic effects of fixed-asset investment trends which can basically reflect the effect of changes in the investment during the period of promoting economic growth mainly by investment. Besides, China's labor productivity has a big gap compared with those of developed countries. In 2001, Chinese total labor productivity is about 1/40 of Japan, 1/12 of South Korea, 1/45 of USA and 1/30 of Germany. Our unit energy production for the GDP is only three-fourths that of Japan and Germany. High input and low output results in absolute waste of funds, materials, and energies. These wastes result in turn in shortages of funds, materials, and energies and lead to a vicious cycle.

12.2 Spatial Layout and Situation of Chinese Industrial Structure

Since reform and opening up, by virtue of the status, strength and advantages of national policies, our southeast coastal provinces have rapidly progressed in foreign economic exchanges and economic developments. But the economic development has lagged far behind in central and western provinces owing to poor investment environment and limited policy liberalization. Our current economic levels of the provinces are uneven because of geographical, cultural, economic, policy and many other reasons. So our country should not be generalized, but local conditions must be considered when carrying out macro-control and constituting industrial policies.

Well-known American economists such as Chenery consider the per capita income level as the classification standard of the various stages of economic development. The World Bank statistics and that of other countries, the basis of inspection system and analysis at different stages of development of the economic structure of the countries all use the per capita income level. The period of the whole change process from undeveloped economies to the mature industrial economies is divided into six stages:

Stage 1: The average national income is between US$ 140 and 280.

Stage 2: The average national income is between US$ 280 and 560.

Stage 3: The average national income is between US$ 560 and 1120.

Stage 4: The average national income is between US$ 1120 and 2100.

Stage 5: The average national income is between US$ 2100 and 3360.

Stage 6: The average national income is between US$ 3360 and 5040.

Stage 1 is the economically backward stage and is before any development. Stage 6 is the developed economy stage and is a symbol of industrial economic maturity. In addition, we use the index called the degree of order to reflect the rational level of industrial structure of various provinces. First, we determine the positions of provinces in the process of industrialization on the basis of Chenery's division of stages of industrial economy and the level of economic development. Then we investigate the adaptation of industrial structure and economic development level of each province using the degree of order. If the degree of order of a certain province is low or high, it indicates that its industrial structure is lagging behind or advanced in the level of economic development. Then we make adjustment for its separate divisions for industrial economic periods of provinces according to the actual backwardness or forwardness situation.

12.2.1 *Classification of Industrial Economic Period in All Provinces*

The division for our industrial economic development period carried out according to Chenery's standard of industrial economic stages is shown in Table 12.5. We see that our economic development of various provinces is extremely unbalanced. The per capita GDP of Shanghai is US$ 4915, which is the highest in all the provinces and the lowest is US$ 381 in Guizhou.

As reflected in the industrialization process, Shanghai and Beijing have stepped into mature industrialization process periods, Tianjing is in the fifth period. Zhejiang, Guangdong, Jiangsu, Fujian, Liaoning, Shandong, and Heilongjiang are in the fourth period, Gansu and Guizhou are still in the second period. The others are in the third period.

12.2.2 *Degrees of Order for Industrial Structures of Provinces*

The degrees of order for industrial structures of our 31 provinces from 2000 to 2003 are shown in Table 12.6.

12.2.3 *Classification of Provinces according to the Process of Industrial Structure Advancement*

On the basis of Chenery's division of industrialization process, we divide the process of industrial structure advancement of each province

Table 12.5 Classification of Per Capita GDP in Provinces and Cities

PROVINCE/CITY	PER CAPITA GDP (US$)[a]	CLASSIFICATION OF INDUSTRIALIZATION PERIOD ACCORDING TO CHENERY
Shanghai	4811	Sixth period: per capita GDP between US$ 3360 and 5040
Beijing	3266	Fifth period: per capita GDP between US$ 2100 and 3360
Tianjing	2612	
Zhejiang	1955	
Guangdong	1770	
Jiangsu	1689	Fourth period: per capita GDP between US$ 1120 and 2100
Fujian	1586	
Liaoning	1527	
Shandong	1370	
Heilongjiang	1200	
Hebei	1078	
Xinjiang	1011	
Hubei	977	
Jilin	972	
Hainan	922	
Neimenggu	856	
Hunan	771	
Chongqing	762	
Qinghai	741	
Henan	733	Third period: per capita GDP between US$ 560 and 1120
Shanxi (Taiyuan)	729	
Xizang	690	
Ningxia	683	
Anhui	676	
Sichuan	667	
Jiangxi	657	
Shanxi (Xian)	652	
Yunnan	615	
Guanxi	606	
Gansu	528	Second period: per capita GDP between US$ 280 and 560
Guizhou	372	

Source: Statistical Yearbook, 2001, 2002, and 2003.

[a] Exchange rate: 1 dollar = 8.27 yuan.

into several types. We define the degree of order of the industrial structure as the threshold value of 0.7. If the degree of order of a certain province is less than 0.7, it indicates that the industry structure is not yet standard. From Table 12.6, we see that for Guizhou, Hainan, and Xizang, the degree of order is less than 0.7, Guizhou is in the

Table 12.6 Degree of Order for Industrial Structures in China

PROVINCE	2000	2001	2002	2003	AVERAGE
Anhui	0.789	0.792	0.792	0.788	0.790
Beijing	0.909	0.912	0.912	0.903	0.909
Chongqing	0.901	0.901	0.896	0.875	0.893
Fujian	0.894	0.882	0.869	0.854	0.875
Gansu	0.728	0.728	0.724	0.712	0.723
Guandong	0.825	0.830	0.829	0.815	0.825
Guangxi	0.868	0.907	0.940	0.899	0.904
Guizhou	0.662	0.698	0.692	0.674	0.682
Hainan	0.526	0.533	0.522	0.515	0.524
Hebei	0.839	0.847	0.847	0.826	0.840
Heilongjiang	0.716	0.724	0.727	0.716	0.721
Henan	0.732	0.736	0.734	0.724	0.732
Hubei	0.820	0.822	0.830	0.845	0.829
Hunan	0.896	0.900	0.894	0.921	0.903
Jiangsu	0.809	0.812	0.808	0.795	0.806
Jiangxi	0.939	0.906	0.850	–	0.898
Jilin	0.885	0.904	0.901	0.875	0.891
Liaoning	0.837	0.855	0.862	0.866	0.855
Neimenggu	0.795	0.853	0.903	0.852	0.851
Ningxia	0.797	0.803	0.794	0.756	0.788
Qinghai	0.826	0.817	0.799	0.779	0.805
Shandong	0.834	0.839	0.832	0.805	0.828
Shanghai	0.841	0.841	0.843	0.822	0.837
Shanxi (Taiyuan)	0.763	0.748	0.745	0.730	0.747
Shanxi (Xian)	0.810	0.793	0.780	0.756	0.785
Sichuan	0.789	0.849	0.836	0.823	0.824
Tianjing	0.833	0.840	0.844	0.828	0.836
Xinjiang	0.884	0.924	0.924	0.857	0.897
Xizang	0.619	0.699	0.681	0.776	0.694
Yunnan	0.759	0.773	0.776	0.769	0.769
Zhejiang	0.800	0.814	0.819	0.810	0.811

Note: For detailed calculation process refer to Section 12.4.

worst stage, Hainan and Xizang are in the third stage. So Hainan and Xizang, and Gansu and Guizhou are in the same type of industrial structure.

Shanghai is in the sixth stage, Beijing and Tianjin are in the fifth stage; they are all in the mature stage, so we place these three cities together. The division is shown in Table 12.7.

Table 12.7 Classification of Provinces in Terms of Per Capita GDP for Industrial Structure

PROVINCE	PER CAPITA GDP (US$)[a]	DEGREE OF ORDER OF INDUSTRIAL STRUCTURE[a]	CLASSIFICATION OF INDUSTRIALIZATION PERIOD ACCORDING TO CHENERY
FIRST TYPE			
Shanghai	4811	0.837	Sixth period: per capita GDP between US$ 3360 and 5040
Beijing	3266	0.909	Fifth period: per capita GDP between US$ 2100 and 3360
Tianjing	2612	0.836	
SECOND TYPE			
Zhejiang	1955	0.811	Fourth period: per capita GDP between US$ 1120 and 2100
Guangdong	1770	0.825	
Jiangsu	1689	0.806	
Fujian	1586	0.875	
Liaoning	1527	0.855	
Shandong	1370	0.828	
Heilongjiang	1200	0.721	
THIRD TYPE			
Hebei	1078	0.840	Third period: per capita GDP between US$ 560 and 1120
Xinjiang	1011	0.897	
Hubei	977	0.829	
Jilin	972	0.891	
Neimenggu	856	0.851	
Hunan	771	0.903	
Chongqing	762	0.893	
Qinghai	741	0.805	
Henna	733	0.732	
Shanxi (Taiyuan)	729	0.747	
Ningxia	683	0.788	
Anhui	676	0.790	
Sichuan	667	0.824	
Jiangxi	657	0.898	
Shanxi (Xian)	652	0.785	
Yunnan	615	0.769	
Guanxi	606	0.904	
FOURTH TYPE			
Hainan	922	0.524	Third period: per capita GDP between US$ 560 and 1120
Xizang	690	0.694	
Ganshu	528	0.723	Second period: per capita GDP between US$ 280 and 560
Guizhou	372	0.682	

[a] Average values from 2001 to 2003.

12.3 Our Target of Industrial Structure Adjustment

Using Leontief's fast track model, we establish the Chinese industrial structure adjustment programs and try to make Chinese economy step into the fast track mode as soon as possible during the 11th Five-Year Plan and rationalize the industrial structure step by step. First, we put the Chinese input–output table of 2000 into 3 × 3 input–output tables as shown in Table 12.8.

We get the direct consumption coefficient matrix A:

$$A = \begin{bmatrix} 0.1526 & 0.0509 & 0.0198 \\ 0.2069 & 0.5662 & 0.2840 \\ 0.0622 & 0.1028 & 0.2044 \end{bmatrix}$$

Then we calculate the coefficient matrix B of fixed assets usage as follows. When calculating matrix B, we define b_{1j}, $b_{3j} = 0$ $(j = 1, 2, 3)$ because the primary industry and the tertiary industry do not form fixed assets. According to Chinese Statistical Yearbook of 2000, the fixed assets of the secondary industry is 85,171 (10^8 RMB). Then we can calculate the fixed assets of the primary industry and the tertiary industry. The average proportion of the fixed assets of the three industries is 1.5%:30.4%:68.1%, so the fixed assets of the primary industry are 4422 (10^8 RMB) and tertiary industry is 191007 (10^8 RMB). Matrix B is as follows:

$$B = \begin{bmatrix} 0 & 0 & 0 \\ 0.1672 & 0.4924 & 3.2856 \\ 0 & 0 & 0 \end{bmatrix}$$

Calculate the consumption coefficient matrix T next. The proportion of the employment structure of the three industries in 2000 was 50.0%:22.5%:27.5%, we can get matrix T as shown next:

$$T = \begin{bmatrix} 0.2071 & 0.0143 & 0.0518 \\ 0.4005 & 0.0276 & 0.1002 \\ 0.2312 & 0.0159 & 0.0579 \end{bmatrix}$$

Table 12.8 Input–Output Table for Three Major Industries in 2000 (10^8 RMB)

INPUT		OUTPUT							
		INTERMEDIATE DEMAND							
		PRIMARY	SECONDARY	TERTIARY	FINAL CONSUMPTION	CAPITAL FORMATION	INTER-REGION NET FLOW	OTHER	TOTAL DEMAND
Intermediate input	Primary	4035.55	8799.43	1148.89	10955.53	1108.80	41.7	358.39	26448.27
	Secondary	5472.57	97931.22	16507.64	21186.06	30445.35	1130.88	295.91	172969.65
	Tertiary	1644.11	17786.33	11880.15	23935.24	945.64	2344.86	–401.47	58134.86
GDP		15296.04	48452.66	28598.17					
Total supply		26448.27	172969.65	58134.86					

The dynamic input–output model is as follows:

$$x(t) = Ax(t) + B[x(t+1) - x(t)] + Tx(t)$$

We suppose the economic growth rate α of each department are the same, we get

$$x(t) = Ax(t) + B[(1+\alpha)x(t) - x(t) + Tx(t)$$
$$= Ax(t) + \alpha Bx(t) + Tx(t)$$

$$\frac{1}{\alpha}x(t) = (I - A - T)^{-1}Bx(t)$$

where $H(I - AT)^{-1}B$. According to Leontief's model of fast track we get average growth rate $\alpha = 14.36\%$, and the best structure of the three industries is 8.16%:61.665%:30.18%.

In fact, the scale of Chinese three industrial structures in 2000 is 10.3%:67.6%:22.1%. So we can see that there's a certain gap between our actually industrial structure and the balanced growth path and it should be adjusted properly to step into the fast track. We plan to use six years time to realize the target of the best structure.

$$x(10) = \begin{bmatrix} 8.16\% \times q \\ 61.66\% \times q \\ 30.18\% \times q \end{bmatrix}$$

q is the coefficient for carrying out total output maximization.

$$\max q$$
$$\begin{cases} Ax(0) + B[x(1) - x(0)] + Tx(0) \le x(0) \\ Ax(1) + B[x(2) - x(1)] + Tx(1) \le x(1) \\ \cdots \\ Ax(9) + B[x(10) - x(9)] + Tx(9) \le x(9) \\ \bar{x} \cdot q \le x(10) \\ x(1), x(2), \ldots, x(10) \ge 0, \quad q \ge 0 \end{cases}$$

$x(10)$ is the three industrial structures in the initial stages. $\hat{x}$is the best industrial structure, $x(1)$, $x(2)$, …, $x(10)$ are the structures from 2001 to 2010.

Table 12.9 Comparison of Simulation Value and Actual Value for Three Major Industries in China (%)

	PRIMARY INDUSTRY		SECONDARY INDUSTRY		TERTIARY INDUSTRY	
YEAR	ACTUAL STRUCTURE	OPTIMIZED STRUCTURE	ACTUAL STRUCTURE	OPTIMIZED STRUCTURE	ACTUAL STRUCTURE	OPTIMIZED STRUCTURE
2001	15.8	15.7	50.1	52.1	34.1	32.2
2002	15.4	15.2	51.1	50.6	33.5	34.2
2003	14.8	14.6	52.9	49.9	32.3	35.5

When we use Lindo software to solve this linear programming model, we find that we should step into fast track in the first year which is impossible. So we consider adding some constraints for controlling this phenomenon. Because the average growth rate of three industries is 14.45%, we believe it is too fast. But as the basic industry of the national economy, the growth rate of the primary industry should not be less than 4%, while the growth rate of the tertiary industry should not be higher than 18%. So we add constraints to analog structure and get the structure of each industry (see Table 12.9).

From Table 12.10, it is seen that the Chinese industrial structure has nearly adjusted to the best one and has made the largest output.

We can get the conversion factors of GDP and social industrial production of the three industries: 0.5879, 0.2893, and 0.4975. Then we can get the adjustment shown in Table 12.11.

We can see that after six years of adjustment, the proportion of the primary industry and secondary industry has decreased gradually while the proportion of tertiary industry has increased.

Table 12.10 Industrial Structure Adjustment Program in Terms of Total Social Productive Value (%)

YEAR	PRIMARY INDUSTRY	SECONDARY INDUSTRY	TERTIARY INDUSTRY
2005	9.45	64.12	26.43
2006	9.16	63.58	27.26
2007	9.06	63.16	27.78
2008	8.93	62.85	28.22
2009	8.33	62.33	29.34
2010	8.16	61.66	30.18

Table 12.11 Industrial Structure Adjustment Program in Terms of GDP (%)

YEAR	PRIMARY INDUSTRY	SECONDARY INDUSTRY	TERTIARY INDUSTRY
2005	14.91	49.79	35.29
2006	14.42	49.26	36.32
2007	14.23	48.83	36.93
2008	14.01	48.52	37.47
2009	13.05	48.05	38.90
2010	12.74	47.38	39.88

12.4 Measurement of the Degree of Order for Chinese Industrial Structure

In order to reflect and analyze the current industrial structure more fully, we take our industrial structure, world standard industrial structure, five-year goal plan of industrial structure, and the best structure of fast track model as the target structure and analyze their degrees of order.

12.4.1 Analysis of the Degree of Order of the Chinese and World Standard Industrial Structure

Taking the standard model of world economic growth and structure changing relationship as the reference, we analyze the status of Chinese industrial structure. According to statistical data from different economic levels in world development report of the World Bank in 2000–2001, we calculate the international standard of average GDP and industrial structure change, then we do research about the relationship between the industrial structure and economic development level. As the average income increase, the added value of the primary industry for the proportion of GDP obviously decreases, while that of the tertiary industry increases.

When using Chanmry's standard of per capita total economic output, we should express the GDP of the countries in dollars. Calculated according to the exchange rate, our per capita GDP was more than US$ 1000 in 2003. So we are at the level of low and middle income countries as far as World Bank report and industry structures are concerned. We may calculate the industry structure goal on the basis of reference to various income levels of the corresponding industrial structures (Table 12.12).

Table 12.12 Comparison of International Per Capita GDP and Industrial Structure (%)

AVERAGE ECONOMIC LEVEL	PER CAPITA GDP IN 1999 (US$)	PRIMARY INDUSTRY	SECONDARY INDUSTRY	TERTIARY INDUSTRY
World	5,057	4	33	63
Low income countries	442	27	30	43
Mid-low income countries	1,230	15	40	46
Middle income countries	2,058	10	36	55
Mid-high income countries	5,092	7	32	61
High income countries	26,560	2	31	67

Note: 1 US$ = 8.27 yuan. Data in this table are calculated according to World Bank Report in 2000 and 2001.

Here we use the method of weighted averages to determine the industrial structure goal of autonomous regions and municipalities nationwide. For example, the per capita GDP of Beijing in 2003 was US$ 31,613; according to the method of weighted averages, the result is US$ 3823, which puts it in the category of middle and middle-high income countries. So we define it as the industrial structure goal. Now we have $3823 \in [2058, 5092]$, so we get the weight

$$\alpha = \frac{5092 - 3823}{5092 - 2058} = 0.41826$$

The proportion of target industries were weighted by means of middle-income developed countries and high-income developed countries in 1999. The target industry structure of Beijing is as follows:

$$\text{Proportion of the primary industry } \alpha \times 10 + (1 - \alpha) \times 7 = 8.3$$

$$\text{Proportion of the secondary industry } \alpha \times 36 + (1 - \alpha) \times 32 = 33.7$$

$$\text{Proportion of the tertiary industry } \alpha \times 55 + (1 - \alpha) \times 61 = 58.5$$

Thus, we get the target industry structure of Beijing as 8.3:33.7:58.5. Likewise, the per capita GDP of China in 2003 was US$ 1092, and the target industry structure was 17.1:38.2:45.5.

The composition of three industries added value is shown in Table 12.13. There is still considerable disparity in our industrial structure compared with the world standard. The main problem is that our primary industry is still weak compared with the secondary industry; its

Table 12.13 Composition of Three Industries Added Values (%)

YEAR	PRIMARY INDUSTRY	SECONDARY INDUSTRY	TERTIARY INDUSTRY
2000	16.4	50.2	33.4
2001	15.8	50.1	34.1
2002	15.4	51.1	33.5
2003	14.8	52.9	32.3

Sourse: *Chinese Statistical Yearbook*, 2003 and *The Statistical Bulletin of the National Economy and Social Development*, 2003.

foundation status is not steady, and the tertiary industry is lagging behind.

The synergy of industrial structure calculated according to the degree of order model is shown in Table 12.14. We see that the synergy of industrial structure does not have an obvious trend. Especially in 2003, it declined to a value below 0.8. The reason may be the decreasing on outputs of the tertiary industry.

12.4.2 Degree of Order Analysis of Our Industrial Structure and the Goal of the 11th Five-Year Plan

The added value structures of the three industries have been adjusted from 16.4:50.2:33.4 (2000) to 14.8:52.9:32.3 (2003). The basic feature is that the proportion of primary industry added value in the GDP drops steadily, the proportion of secondary industrial added value in the GDP rises with some fluctuation, and the proportion of tertiary industry is quite stable. During the 10th five-year plan period, the anticipated target of our country's readjustment of the economic structure was 2005, and the structure of three industries added value was 13:51:36. Compared with actual structure of 2000, during the 10th five-year plan period, the proportion of the primary industry added value in the GDP should drop by 0.68% every year, the secondary industry added value proportion should rise by 0.16%,

Table 12.14 Comparison of Relative Degree of Order

YEAR	DEGREE OF ORDER
2000	0.811
2001	0.814
2002	0.805
2003	0.788

Table 12.15 Comparison between the GDP Structure of Three Major Industries and Annual Goal from 2000 to 2003

YEAR	PRIMARY INDUSTRY	SECONDARY INDUSTRY	TERTIARY INDUSTRY
ACTUAL VALUE			
2001	16.4	50.2	33.4
2002	15.8	50.1	34.1
2003	15.4	51.1	33.5
2005	14.8	52.9	32.3
GOAL OF THE 10TH FIVE-YEAR PLAN			
2001	15.72	50.36	33.92
2002	15.04	50.52	34.44
2003	14.36	50.68	34.96
2005	13.0	51.0	36.0

Source: The 10th Five-Year Plan. *Chinese Statistical Yearbook*, 2003 and *The Statistical Bulletin of National Economy and Social Development*, 2003.

the tertiary industry added value should rise by 0.52%. According to the decomposition year adjustment goal (Table 12.15), in 2003, added value structure of the three industries in our country should have been 14.36:50.68:34.96. So, the development of the primary industry is basically consistent with the goal of the 10th five-year plan; the speed could have been more. The tertiary industry still has disparity and the proportion of the secondary industry is higher than the planned level.

We take the industrial structure target value of the 10th five-year plan to calculate the degree of order in 2003 and compare with that of the 10th five-year plan (Table 12.16). We see that the degree of order is rising steadily, which indicates that the adjustment of industrial structure has conformed to the plan and moved forward continuously.

Table 12.16 Degree of Order of Industrial Structure Compared to That of the 10th Five-Year Plan

YEAR	DEGREE OF ORDER
2000	0.928
2001	0.940
2002	0.953
2003	0.972

Table 12.17 Degree of Order of Actual Structure and Optimal Structure

YEAR	DEGREE OF ORDER
2000	0.829
2001	0.838
2002	0.848
2003	0.864

12.4.3 Degree of Order Analysis of Actual Structure and Optimal Structure

We get the optimal structure according to the fast track model: 12.74:47.38:39.88. Taking the optimal model as target structure, we get the degree of order of actual structure and optimal structure from 2000 to 2003 (Table 12.17). We see that the degree of order of actual structure and optimal structure is gradually increasing from 2000 to 2003. The achievement of adjusting the industrial structure is obvious.

12.4.4 Comparison of the Degree of Order of Industrial Structure in Different Provinces

According to the per capita GDP of all provinces and the world standard, we calculate the target value of the corresponding industrial structure, then calculate, analyze and compare the degree of order of industrial structures of all provinces to the world standard from 2000 to 2003.

We divide the provinces into three types according to the per capita GDP and the target industrial structure. The first type includes Beijing, Tianjin, and Shanghai, belonging to the developed regions, with high urbanization degree, whose per capita GDP was US$ 3823, 3129, and 5633, respectively; the second type includes Liaoning, Jiangsu, Zhejiang, Fujian, Shandong, and Guangdong, belonging to the relatively developed regions, whose per capita GDP is more than US$ 1500; the rest of the provinces, belonging to less developed regions are included in the third type, whose per capita GDP is less than US$ 1500, sometimes even less than US$ 1000.

First, we define the object structure and calculate the industrial structure target values with the help of world standard and the per capita GDP of different provinces in 2003. The results are shown in Table 12.18.

Table 12.18 Three Industries Target Structure of All Provinces in 2003 (%)

PROVINCE	PER CAPITA GDP (US$)[a]	PRIMARY INDUSTRY	SECONDARY INDUSTRY	TERTIARY INDUSTRY
FIRST TYPE				
Shanghai	5633	6.9	32.0	61.2
Beijing	3823	8.3	33.7	58.5
Tianjing	3129	8.9	34.6	57.1
SECOND TYPE				
Zhejiang	2386	9.7	35.6	55.6
Shandong	1651	12.5	38.0	50.6
Liaoning	1724	12.0	37.6	51.4
Jiangsu	2031	10.2	36.1	54.7
Guangdong	2045	10.1	36.1	54.9
Fujian	1815	11.5	37.2	52.4
THIRD TYPE				
Heilongjiang	1405	16.9	33.6	50.2
Hebei	1271	14.8	39.8	46.4
Xinjiang	1171	15.9	39.3	45.8
Jilin	1128	16.6	38.7	45.6
Hubei	1087	17.2	38.2	45.5
Neimenggu	1055	17.7	37.8	45.3
Hainan	1046	17.8	37.7	45.3
Chongqing	972	18.9	36.7	45.0
Shanxi (Taiyuan)	892	20.1	35.7	44.7
Qinghai	880	20.3	35.6	44.7
Hunan	876	20.4	35.5	44.7
Xizang	831	21.1	34.9	44.5
Ningxia	802	21.5	34.6	44.4
Shanxi (Xian)	784	21.8	34.3	44.3
Anhui	781	21.8	34.3	44.3
Henna	778	21.9	34.3	44.3
Sichuan	758	22.2	34.0	44.2
Guangxi	721	22.8	33.5	44.1
Jiangxi	704	23.0	33.3	44.0
Yunnan	683	23.3	33.1	43.9
Ganshu	603	24.5	32.0	43.6
Guizhou	435	27.1	29.9	43.0

Note: The data are from the Statistics Bulletin of the provinces in 2003.

[a] Exchange rate: 1 US$ = 8.27 yuan.

From the relevant statistical yearbooks, we get the composition of three industries structures added value of all provinces from 2000 to 2003. We get the degree of order of all provinces by the degree of order model (Table 12.19).

The degrees of order of all provinces from 2000 to 2003 are gradually increasing, which indicates that the goal of industrial structure adjustment is correct. Further, the proportion of the tertiary industry declined in 2003 because of severe acute respiratory syndrome (SARS), the degrees of order of some provinces declined in 2003. However, the range of increase is small and has fluctuations, indicating that the adjustment of industrial structure is a gradual and long-term process.

In the computational process of industrial structure degree of order, the target structures of all provinces are based on their own developmental levels, and the object constructions are not the same. Therefore, the results of different provinces do not have commensurability. Each value reflects the rationalization degree and the changing tendency of industrial structure. Some provinces have high degrees of order, which indicates that the current industrial structure and the economic development level are well adapted.

The industrial structure degree of order of Beijing is the highest between Beijing, Tianjin, and Shanghai. In fact, by comparing the results between the actual industrial structure and the target structure in Beijing, the proportion of the primary industry is seen to be lower than that of the target structure, the proportion of secondary industry are almost the same; the proportion of the tertiary industry is higher than that of the target structure; in Tianjin and the Shanghai, the proportions of the secondary industry are higher and that of the primary industry and tertiary industry lower than those of the target structure.

The proportions of the secondary industry of all six provinces in the second type are higher than those of the target structure, the proportions of the tertiary industry are lower than that of the target structure, and the proportions of the primary industry are higher except for Fujian; in the case of the rest of the five provinces, they are lower than that of the target structure.

Among the 22 provinces in the third type, the proportions of most provinces of the primary industry and the secondary industry are higher than those of the target structure, the proportions of the

Table 12.19 Degree of Order of Industrial Structures of All Provinces

PROVINCE	2000	2001	2002	2003
FIRST TYPE				
Beijing	0.909	0.912	0.912	0.903
Shanghai	0.841	0.841	0.843	0.822
Tianjing	0.833	0.840	0.844	0.828
SECOND TYPE				
Fujian	0.894	0.882	0.869	0.854
Guangdong	0.825	0.830	0.829	0.815
Jiangsu	0.809	0.812	0.808	0.795
Liaoning	0.837	0.855	0.862	0.866
Shandong	0.834	0.839	0.832	0.805
Zhejiang	0.800	0.814	0.819	0.810
THIRD TYPE				
Anhui	0.789	0.792	0.792	0.788
Chongqing	0.901	0.901	0.896	0.875
Ganshu	0.728	0.728	0.724	0.712
Guangxi	0.868	0.907	0.940	0.899
Guizhou	0.662	0.698	0.692	0.674
Hainan	0.526	0.533	0.522	0.515
Hebei	0.839	0.847	0.847	0.826
Heilongjiang	0.716	0.724	0.727	0.716
Henna	0.732	0.736	0.734	0.724
Hubei	0.820	0.822	0.830	0.845
Hunan	0.896	0.900	0.894	0.921
Jiangxi	0.939	0.906	0.850	–
Jilin	0.885	0.904	0.901	0.875
Neimenggu	0.795	0.853	0.903	0.852
Ningxia	0.797	0.803	0.794	0.756
Qinghai	0.826	0.817	0.799	0.779
Shanxi (Taiyuan)	0.763	0.748	0.745	0.730
Shanxi (Xian)	0.810	0.793	0.780	0.756
Sichuan	0.789	0.849	0.836	0.823
Xinjiang	0.884	0.924	0.924	0.857
Xizang	0.619	0.699	0.681	0.776
Yunnan	0.759	0.773	0.776	0.769

tertiary industry are lower, but the disparity is not very large. The degrees of order of some provinces such as Chongqing and Jilin are high; their actual structures and the target structure are very close. Although the proportions of the tertiary industry are high in Xizang and Qinghai, their primary industries and secondary industries are not particularly developed. And they also have a long way to go for the industrial structure upgrading in the true sense.

Based on the above facts, we believe that in our country the industrial structure upgrading generally is not wise. The best choice is using different strategies according to the actual situation. For example, the tertiary industry of Beijing accounted for 61% in 2002, which was not suitable to overemphasize the development of the tertiary industry. For Tianjin, Shanghai, and six provinces in the second type, we must keep the development of the primary and the secondary industries, then advance the tertiary industry with emphasis, realize the industrial structure upgrading through increasing the proportion of tertiary industry in the GDP. For the 22 provinces in the third type, the primary mission will be promoting the agricultural foundation, advancing the development of the secondary industries and speeding up the industrialization process. These provinces should take development as the first matter, and should not lay emphasis on adjusting industrial structure.

12.5 Choice of Methods to Upgrade Industrial Structure in China

12.5.1 Positively Participating in the World Economy, "Bringing In" and "Going Out" Are Inevitable Choices to Speed Up Industrial Structure Advancement (Way I)

As the globalization of economy cannot be stopped, the influence of the choice and progress of industrial structure upgrading are as follows. First, the arrival of high tide of the world's new scientific and technological revolution, the "information superhighway," has provided all the countries and regions of the world the means and the possibility of choosing the way for "bringing in" and "going out" to reach industry structure advancement. At the same time, it also accelerates the process of advancement of the world's industry structure. Especially, the large popularity of information technology has greatly promoted the development of the world's productive forces, further deepening the international division of labor and more development of the division within the industry, producing a strong global demand for global economic cooperation for the rational allocation of production factors.

Second, the globalization of economic demands needs to establish a relatively perfect market economy system for countries of the world. This makes the less-developed countries choose the radical reform of replacing the old imperfect market economic system and with a new market economy system suited to the national conditions during the process of industry structure advancement. Developing the market economy requires that it be in line with the world market and gradually integrate with the world economy.

Third, the development of the international companies promote the world's transfer of production and technology and provide a "convenient door" for "bringing in" and industry structure upgrading. At the same time, it also places a lot of obstacles for the "going out" of the undeveloped countries. The international companies had developed rapidly after the 1980s, which became the main actors in the global economic cooperation. The trade with other economic bodies has accounted for two-thirds of the world's trade amount. International companies operate and control 80% of the global trade and 90% of technology research, development, and transfer.

Fourth, the international regional economic integration make the "bringing in" and "going out" way of industry structure upgrading an inevitable choice. At the end of the 1980s, the trend of the regional economic integration strengthened significantly and new regional economic integration organizations emerged. Especially the formation and development of the three major regional economic organizations (EU, NAFTA, and APEC) have pointed out the direction for global economic cooperation. According to incomplete statistics, there are more than 30 regional economic organizations in the world at present, in which EU NAFTA, and APEC take up 80% of the GDP of the world and more than 80% of international trade, which greatly have pushed forward the free movement of products, capital, information, and technology within the region, and has made global economic cooperation go forward by eliminating trade and investment barriers and cooperating with fair and mutual benefit.

Joining the WTO accelerates the integration of China into the globalization process. In the future, the possible use of resources, competition, and challenges we face in the process of development of our economy and upgrading of the industry structure will be from the whole world. In the context of the globalization, joining the world's economic circle actively, going along the way of "bringing in" and "going out" to make the industry structure upgrading become our inevitable choice.

12.5.2 Breakthrough in the Region, Spreading, Equilibrium of the Whole Body and Rising Up, a Feasible Regional Choice to the Industry Structure Upgrading (Way II)

12.5.2.1 Main Idea of Breakthrough in the Region, Spreading, Equilibrium of the Whole Body, and Rising Up China's economic disparities between regions are large. According to statistics of 2002, the per capita GDP in Shanghai, which is the most economically developed area of China, has reached US$ 4945, which stays in the sixth stage in Chenery's industrialization stage classification of economic development. However, in the backward area such as Guizhou, the per capita GDP is only US$ 381, which is in second stage in Chenery's theory. China's regional economic development covers five stages of the total six stages in Chenery's theory. Facing the huge difference of China's unbalanced regional economic development, the method of choice in upgrading the industry structure of China plays a significant role in accelerating the process.

According to the history of world economic development and upgrading of industry structure and China's actual situation, from the point of view of economic spreading, we should choose as follows: from the priority of individual points to small regional area, then to large regional area, then pan-regional final leap toward the overall balance of the country (see Figure 12.2).

The regions such as Shanghai, Beijing and Tianjin, due to they are economically developed, the completementary of market structure and industry structure advancement, these regions are very easy to accept, create and deliver the advanced productive force. They become the vanguard of becoming the advancement of the region industry structure. They will first get the breakthrough in the industry structurc upgrading, and thus to promote and stimulate their wider regions and the escalation of the regional industrial structure.

12.5.2.2 Concentration and Spread of Yangtze River Delta Economic Circle for Industrial Structure Upgrading China's entire industrial structure is to promote the process of upgrading, in general, which

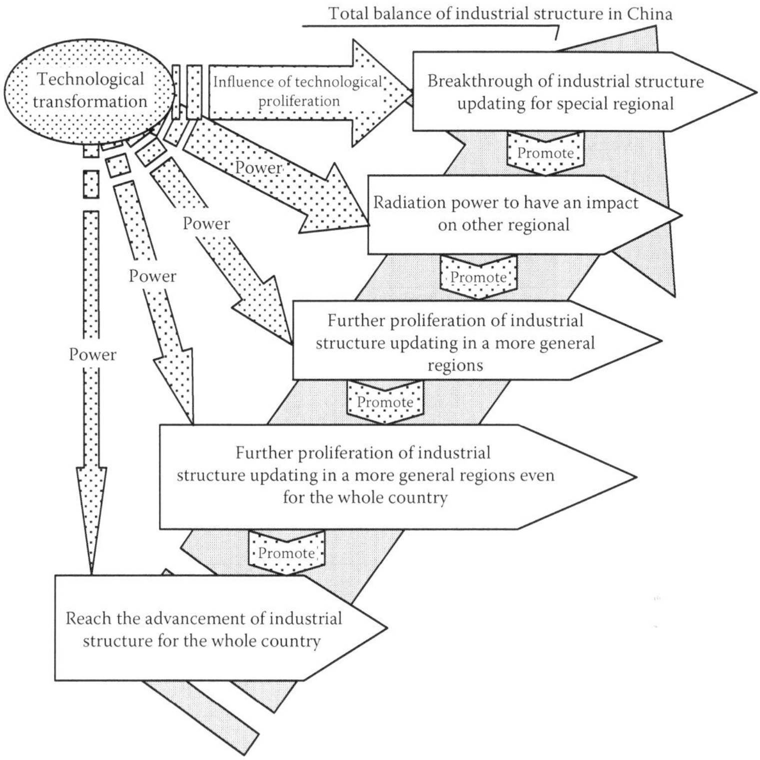

Figure 12.2 A schematic of the regional breakthrough proliferation, overall balance, jump, and high grade of the way of China's industrial structure.

includes a rail link from east to west, from the coastal areas to the areas under development. The Yangtze River Delta region crosses Shanghai, Jiangsu, and Zhejiang Provinces. Shanghai, as the center of the Yangtze River Delta cities, has become the leader of Yangtze River Delta economic circle, the Pan-Yangtze River Delta economic circle, and the whole Yangtze River basin, and a leader even in China's upgrading of the center of the industrial structure and economic development.

In 2002, Shanghai provided and created the 5.2% of the GDP, 11.9% fiscal revenues and 9.8% import and export with 1% of the population in national economy, which is less than 0.2% of national land area. At the same, Shanghai accounted for only 17.6% and 6.4% of population and land area of the Yangtze River Delta region, while the GDP, fiscal revenue, foreign trade import and export volume in the Yangtze River Delta region accounted for 28.3%, 54.4%, and 34.6%, respectively. The rapid economic development in the south of Jiangsu, especially Suzhou–Wuxi–Changzhou is affected by the level of Shanghai's economic gradient, largely due to the economic radiation of Shanghai. Located within a short distance of Shanghai, the cities of Shanghai–Nanjing line such as Suzhou, Wuxi, Changzhou, and Zhenjiang reflect the total economic output and the financial strength of the local export-oriented economy in particular.

In the 21st century, Shanghai will become the sixth-largest metropolitan area in the world after New York, Toronto and Chicago, Tokyo, Paris and Amsterdam, London and Manchester. The planning will be centered in Shanghai and cover 15 cities of the Yangtze River Delta, major regional transportation systems, comprehensive improvement of ecological environment, the formation of a unified market and service system for financing the establishment of cooperation. At present, the Yangtze River Delta has become one of the first choice of the international capital in China. Data have shown that China's actual foreign investment is in place, the investment in the Yangtze River Delta is more than 30%. The rapid accumulation of capital is anxiously waiting for the industry to integrate the region as a whole in order to avoid redundant construction and vicious competition. Among 15 cities in the Yangtze River Delta, there are 19 state-level development zones and more than 100 provincial development zones. There are more than 100 berths which have been built or to be built in the section of Yangtze River below Nanjing.

A powerful radiation is formed in the entire Yangtze River Delta economic circle, which is driven by Shanghai. The radiation power can be extended to the nation and the world, then promote the optimization and upgrading of industrial structure in our country.

12.5.2.3 Concentration and Spread of Pearl River Delta Economic Circle for Industrial Structure Upgrading The Pearl River Delta region includes Guangdong Province, the south-central and the lower reaches of the vast region of Pearl River estuary. It centers around the city of Guangzhou, south of which is Shenzhen, Zhuhai Special Economic Zone; Shantou is on the east, Zhanjiang is on the west, which forms a fan of opening up at different levels. After China entered the WTO, the "Greater Pearl River Delta" economic circle with Guangdong, Hong Kong, and Macao, three regions constituting the fulcrum, is accelerating the formation of the three economic regions integration; cooperation and upgrading of the three regions are speeding up.

After entering the new century, Guangdong put forward the idea of Pan-Pearl River Delta region on its own need and demand of national development, referred to as "9 + 2," namely: Guangdong, Fujian, Jiangxi, Guangxi, Hainan, Hunan, Sichuan, Yunnan, Guizhou, coupled with the formation of Hong Kong and Macau regional economic circle. Guangdong Province will play a leading role of cooperation to promote the economic development.

Guangdong and its neighboring Pearl River in upper reaches of the eight provinces and autonomous regions (Fujian, Hainan, Jiangxi, Hunan, Guangxi, Guizhou, Yunnan, and Sichuan) have a close working relationship in transportation, logistics management, energy development, scientific research and technology, and cultural tourism. According to incomplete statistics, since the "Ninth Five-Year Plan" period, the eight provinces and Guangdong signed over 8000 economic and technical cooperation projects and reached a total amount of 550 billion yuan. This region has a large population, economic output, and a larger potential market; in 2002, the region reached 3.4474 trillion yuan of GDP (accounting for 33.67% of the nation). The region has a very important position in future economic development. As a result of economic geography, natural conditions, features, relative advantages, if we combine the economy of the region with that of Guangdong, Hong Kong, and regional economic development strategies and establish "9 + 2" Pan-Pearl River Delta economic cooperation zone, then we will build an economic system leading the development of south China and having an impact on the economic development of countries in southeast Asia.

The Pearl River Delta Region has played a tremendous role in China's reform and opening up, economic and social development, and the process of industrial structure upgrading. We believe that with the realization of the concept of Pan-Pearl River Delta, it will have significant meaning to China's and the world's economic development and industrial strucure upgrading.

12.5.2.4 Concentration and Spread of the Capital Economic Circle Industrial Structure Upgrading On May 18, 2004, officials and experts from Beijing, Tianjin, and Hebei got together to discuss the metropolitan area and the capital of Beijing–Tianjin–Hebei economic zone plans for the future. For the first time they put forward the "3 + 2" plan, as well as the capital economic circle "one axis, two cores, and three districts" as the framework for the development of the strategic concept of metropolitan Beijing, Tianjin, and Hebei. The so-called "3 + 2" is on the basis of existing cooperation of Beijing–Tianjin–Hebei, Inner Mongolia, and Shandong, adding parts to play the role of meat custard in the labor supply side advantages, technology, information, personnel, resources, and market in a larger space within the flow and configuration in Inner Mongolia and Shandong.

The so-called "one axis" is the Beijing–Tianjin–Tanggu Expressway, with the formation of high-tech industries, which will be built here together with their own interests of the industry chain and industry clusters. "Two cores" describe Beijing and Tianjin as a dual-core of the capital economic circle, which will combine the strengths of capital in Beijing and advantages of the port of Tianjin. Beijing and Tianjin have the advantages of knowledge-based economy to promote the formation of metropolitan Beijing, Tianjin and ration the geographical division of labor. The "three areas" signify Beijing–Tianjin–Tangshan industry area; Beijing and Tianjin are insurance industry areas, and Jing Zhang is the ecological conservation area. Beijing–Tianjin–Tangshan industrial zone will be developed into an energy and raw materials supply based in Beijing, Tianjin, and Hebei and resource-intensive manufacturing base; Beijing–Tianjin area security industry will build a petrochemical downstream industries, urban-oriented industries such as light industry processing characteristics set group to form Beijing and Tianjin's modern manufacturing and processing bases and complementary parts; Jing Zhang ecological conservation area will take tourism and leisure to be a pillar industry and develop into the tourist base of Beijing, Tianjin, and Hebei metropolitan area, green production and processing base of agricultural and livestock products, electricity, and other clean energy supply base.

Data indicate that in the five provinces (municipalities) of capital economic circle, the GDP, the per capita GDP, and the difference of three industries are relatively large according to the Chenery's theory; five provinces (municipalities) cover four stages of the six stages. This shows that Beijing is the center of this economic circle and Tianjin a second center; the productivity differences of the layout is large, which creates a broad space for upgrading the central industrial cities, core accumulation, and proliferation of potential energy. At the same time, the economic zone is located in the northeast and central regions of the junction, which has great significance for the formation and development of industrial structure due to its high potential for accumulation and spread of the industrial structure upgrading for China, jumping the overall breakthrough.

12.5.2.5 Progress of Realization of China's Industrial Structure Upgrading From a strategic point of view, the progress of China's industrial structure trend is as follows. In the wave of globalization, eastern China has gathered some of the high grade potential of the industrial structure and formed a good tendency of diffusing to the central, western and the northeastern regions. And the implementation and acceleration of development of the western region, the revitalization of old industrial base of the northeast, have well prepared to accept the technology transfer of world industry. In terms of means of the industrial structure upgrading, we should do the following:

First, form a breakthrough point in a number of important economic region core cities of industrial structure upgrading.

Second, based on breakthrough point, form a smaller breakthrough in the region for the regional industrial potential for advancement.

Third, to expand the results, actively promote the industrial structure of the pan-regional spread of advancement.

Fourth, in some less developed regions of productivity and low industrial structure regions, take certain "activation" measures to better coordinate the industrial structure of high-potential areas of high grade energy radiation.

Fifth, grasp the good pattern of the national economic game and world economic game during the process of China's industrial structure upgrading; accelerate the balance and jump of the industrial structure upgrading of China.

12.5.3 Tracked by the Leading Industry, from Low Level to High Level: A Feasible Choice of Industrial Structure Upgrading (Way III)

12.5.3.1 Level Analysis of the Industrial Structure of Different Regions in Our Country On the basis of the extent of industrial structure upgrading, the world's industrial structure can be divided into five levels: the first level is that of the high-tech economy dominated by the contribution rate; the main driving force of economic development is technological innovation, and it has a large number of core technologies; the second level is the financial capital in the economy dominated by the contribution rate; the third is the equipment manufacturing industry as a pillar of economic and social development of the industry; the fourth is the general manufacturing industry to promote economic and social development which has become the main force; the fifth is the energy and raw materials production and processing industry which are the major force for economic and social development.

In fact, as the industrial level and the process of economic and social development are closely related, we use the per capita GDP and industrial structure in an orderly model of the industrial structure of China's various provinces and cities in the delineation of the extent of upgrading (Section 12.2). Compare the results of this division with the Chenery's industrial stage of economic development and industrial division of contrasting levels, which is shown in Table 12.20. This comparison shows that the level of the industrial structure of China's Provinces and cities has a large difference and covers all levels and is low generally. Currently, China's overall industrial structure is in the fourth level. However, with the concentration of industrial layout of China's industrialization process, it will certainly bring the climax of urbanization. Urbanization can increase the investment in infrastructure and increase the demand to the equipment manufacturing industry, which may promote the fourth level of our country to the third level, or the fourth level, with further development, may be skipped to reach the second level.

At the same time, in different levels of regional industrial structure, leading industries of promoting economic and social development are different. How to select the appropriate industry-led regional industrial structure based on the different levels is important to promote the regional and national industrial structure upgrading.

Table 12.20 Division of Provincial, Municipal, and Regional Economic Development, Industrial Level

PROVINCE	PER CAPITA GDP (US$)	LEVEL OF INDUSTRIAL STRUCTURE	CLASSIFICATION OF INDUSTRIALIZATION PERIOD ACCORDING TO CHENERY
FIRST TYPE			
Shanghai	4915	Junction between the first and the second levels; mainly in the second level	Sixth period: per capita GDP between US$ 3360 and 5040
Beijing	3440		
Tianjing	2706		Fifth period: per capita GDP between US$ 2100 and 3360
SECOND TYPE			
Zhejiang	2036	The second, third, and fourth levels; mainly in the third and fourth levels	Fourth period: per capita GDP between US$ 1120 and 2100
Guangdong	1817		
Jiangsu	1740		
Fujian	1632		
Liaoning	1570		
Shandong	1408		
Heilongjiang	1231		
THIRD TYPE			
Hebei	1102	Mainly in the fourth and the fifth levels	Third period: per capita GDP between US$ 560 and 1120
Xinjiang	1014		
Jilin	1008		
Hubei	1006		
Neimenggu	876		
Hunan	794		
Henna	778		
Qinghai	777		
Chongqing	767		
Shanxi	743		
Jiangxi	705		
Anhui	703		
Ningxia	701		
Sichuan	697		
Shanxi	668		
Yunnan	626		
Guangxi	617		

continued

Table 12.20 (continued) Division of Provincial, Municipal, and Regional Economic Development, Industrial Level

PROVINCE	PER CAPITA GDP (US$)	LEVEL OF INDUSTRIAL STRUCTURE	CLASSIFICATION OF INDUSTRIALIZATION PERIOD ACCORDING TO CHENERY
FOURTH TYPE			
Hainan	944	Mainly in the fifth level	Third period: per capita GDP between US$ 560 and 1120
Xizang	737		
Ganshu	543		Second period: per capita GDP between US$ 280 and 560
Guizhou	381		

12.5.3.2 Selection and Development of Regional Leading Industries Upgrading At present, China is in the middle stage of industrialization and has a solid industrial base. A number of emerging industries such as electronics, bioengineering, new materials, and new energy sources have a fixed scale. But there are too many low-tech manufacturing industries in China's total processing industry; a large number of core technologies critically depend on imports; high-tech industries have not yet been formed on a large scale and industry links are lower; the development of downstream industries have yet to be strengthened. At present, China's leading industries should be identified as electronic information, petrochemicals, automobile manufacturing, and construction sectors. Now we mainly discuss the choice and its development of structure of our region at different levels.

1. *Shanghai-centered Yangtze River Delta.* After 50 years of development, this region has developed infrastructure and relatively complete industrial systems, quite a solid scientific and technological strength, and well developed urban communities. The region should focus on high-tech industry, electronic information industry, equipment manufacturing industry, automobile manufacturing, and construction industry development. With Shanghai as the leader, Shanghai, Nanjing, Hangzhou, and a number of small and medium-sized cities as supporting cities, the majority of the hinterland spread out to the Yangtze River to form a composite band-based economy. Shanghai can give priority to the development of capital-intensive and

technology-intensive industries of commercial trade, finance and insurance, real estate, and information and promote the surrounding area for spreading related development of downstream industries; vigorously develop tourism and strengthen environmental governance and pollution remediation.

2. *Pan-Pearl River Delta region with Guangzhou, Shenzhen, Hong Kong, and Macao being the centers.* Pan-Pearl River Delta region is with Hong Kong and Macao as a gateway to the world at the forefront of China's reform and opening up, the location of which is important. The region is the main industrial base of China, and it is at the forefront of reform and opening up. We should strengthen the concept of technology on the basis of industry upgrading and increasing the technological content. We can use the foreign capital to gradually establish technology-intensive and knowledge-intensive industries. We must continue to develop export-oriented economy, carry out economic cooperation with foreign countries, expand opening up, open up international markets and do a good job of economic convergence with Hong Kong and Macao.
3. *Capital economic circle with Beijing being the center.* The automotive, electronic information, bioengineering, new energy, education of this region are developed. Generally, the level of resources is relatively high. The structures of complete and complementary resources in the region are the main features of economic development. The region should be based on electronic information, automobile and petrochemical-based industries and positivily develop bioengineering, new materials, new medicines, and photoelectric machine industries to promote industrial modernization and make full use of technology education advantages of Beijing and Tianjin and speed up the direction of technology-intensive to knowledge development.
4. *Central regions, including Shanxi, Henan, Inner Mongolia, Henan, Hubei, Hunan, Jiangxi, and Anhui.* The central region is in the hub zone of our country. The region has abundant mineral and freshwater resources and is the main supplier of energy and raw materials. The central region should enhance

the transformation and upgrading of traditional industries. The region should accelerate the development of regions along the Longhai Railway, Beijing–Kowloon, Beijing–Guangzhou railway line, create the advantage of convenient transportation, and play the role of linking the east and west. Actively develop a diversified agriculture and make the region along the river and lake plain into the country's major agricultural production base.

5. *Development of northeast region.* The region should expand the automotive and petrochemical industries, with food production industry as pillar, and operate agriculture as industry. Develop toward specialization, commercialization, and modernization of major agricultural industry, become the important base for food, timber, sugar beet, animal husbandry production. Actively develop export-oriented economy, vigorously push forward the regional economic development, and actively participate in northeast Asia, international cooperation, and become the forefront region in getting into Northeast Asian Economic Zone. Speed up the transformation of old industrial bases and improve scale of the economy of the technological level of the automotive, petrochemical, machinery, steel, shipbuilding. Forestry production should achieve a balance between planting and cutting down.
6. *Western regions, including Shanxi, Sichuan, Chongqing, Ningxia, Qinghai and Xizang Provinces.* The western region should promote the development of industrial structure adjustment down-to-earth and step-by-step. Focus on huge investments in infrastructure of western counties, and develop the comparative advantage of the characteristics of the economic and competitive industries or products. Actively guide the eastern part to capital, technology, personnel, and other resources to flow to the central and western regions. Make full use of resources to build an energy and raw materials production base with hydropower development and oil development as a leader, nonferrous metal smelting and petroleum chemical industry, focusing on chemical salts, machine building, and other related processing industries.

12.5.4 From Cities to Rural Areas, the City Leads the Rural Areas: An Inevitable Choice of Advancement Solution Based on the Industrial Structure in Solving the Problem of Rural Areas (Way IV)

Modern economic theory and practice have proven that advanced productive forces often start in cities, especially those which have talent-rich, a complete science and technology education system and developed productive forces. Industrial structure upgrade is often completed in the city first, and then by the city through the integration and transfer of technology to stimulate the upgrading of industrial structure of agriculture, as shown in Figure 12.3. For example, by combining biology, information technology and agriculture to form the biological and agricultural information and take the path of industrialization of agriculture and so on.

China is a large agricultural country, the rural population accounting for 68%. China's total population is 1.3 billion. Rural population is 800 million. The proportion of China's labor force to the total population is only 50%, while the population which is engaged in the primary industry is more than one half. "Rural, agriculture, and farmers" issue have long been key issues of China's economic and social development and industrial structure upgrading. Solve the "three rural" issues, and promote China's industrial structure upgrading, mainly by urbanization, vigorously promoting and accelerating the transfer of the city's advanced productive forces to rural areas, implement industrialized operation of agriculture, reduce China's rural population, promote rural population transfer to cities and towns, vigorously develop the tertiary sector to absorb surplus rural labor force. To achieve the optimization and upgrading of industrial structure in rural areas and promote China's

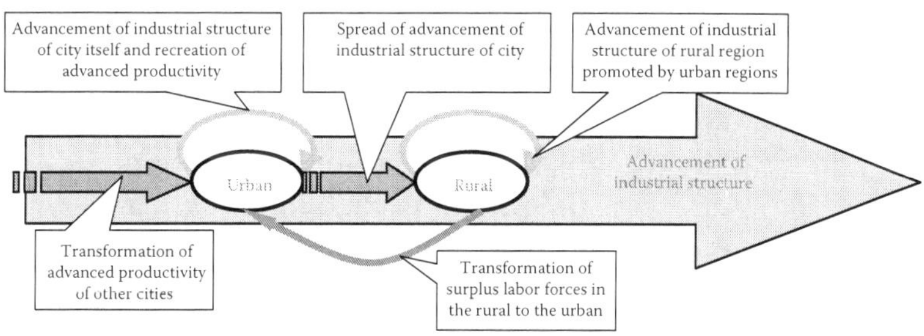

Figure 12.3 Way of solving the problem of industrial structure advancement in rural regions.

urbanization process, the key is to solve the "three rural," problem, agricultural industrialization, urbanization, and farmers in rural areas.

12.5.4.1 Vigorous Promotion of Industrialized Operation of Agriculture, Raising Agricultural Disaster Mitigation, Disaster Resilience, and Consolidation of the Status of Agriculture as Foundation Although the proportion of the primary industry is large, the fundamentals are still weak and the ability to resist natural disasters is also very weak; this has now resulted in a national economy with a fundamental structural contradiction. To follow the path of agricultural industrialization; the key is to enhance agricultural disaster mitigation, disaster resilience, and consolidate the status of agriculture as the foundation. To do a good job in industrialized operation of agriculture, we must do as follows:

First, it is necessary to establish a sound land-use system to achieve the transfer of land use rights in accordance with the law. It is possible to do the cultivation of large centralized land, easy to form specialized agricultural production enterprises, and the formation of large-scale and industrialized operation. Farmers who transfer land usage rights can make use of transfer payments or compensation to engage in the secondary and the tertiary industries, or continue to engage in agricultural production, have access to agricultural enterprises and become industrial workers in agriculture.

The second is to establish a diversified investment mechanism and a reasonable policy of agricultural subsidies. The government should formulate preferential financing, investment policies, diversification of the investment system as soon as possible. It is necessary to form private investment as the mainstay of investment for the state's financial guidance, credit funds for the support of foreign investment, private capital funds and capital market funds and other multi-inputs of agricultural investment and financing system. Provide the industrialization of agriculture leading enterprises benefits such as loan discount incentives to encourage businesses to become bigger and stronger, support the agricultural industrialization leading enterprises, agricultural infrastructure, and scientific and technological innovation projects, encourage the industrialization of agriculture to play the comparative advantages of leading enterprises to participate in international competition. Favor the major grain-producing areas through loans, adjust the credit structure, optimize the investment loan, and gradually make use of the loans for high-yielding, high-quality, and efficient enterprises development.

Third, strong support and development of the regional grain industries, of leading grain-producing areas, and of characteristic features of the areas can help farmers to increase their income. This can be conducive to the production, circulation, and market mechanisms for the rational distribution of agricultural production, to increase the income and improve the capacity to achieve positive interaction.

12.5.4.2 Speeding Up the Process of Urbanization in Rural Areas as Objective Needs of Industrial Structure Upgrading Rural urbanization is the objective requirement of China's industrial structure upgrading, which is the inevitable result of social development. Rural urbanization can contribute to the development of the secondary and tertiary industries, and have great significance to solve the employment problems among surplus rural labor force.

Urbanization is a major strategy in rural areas to bring about economic and social development. There are still more than 19,000 small towns with a population of 170 million. In addition, there are still nearly 30,000 towns, which have a population of more than 50 million. Speeding up urbanization of rural areas will inject new vigor and vitality in our country's economic and social development. Seize the characteristics of the industry. Urban construction has been a strong support for the industry, so that agglomeration effect emerges in urban areas; strengthen the infrastructure, improve the social services and accommodation services, and create the advanced culture of the cities and towns. We need to pay attention to the following principles in the process of urbanization in rural areas.

1. *The principle of industry support:* In the construction of some cities and towns, the driving force of the government is too much and the market forces too weak, which results in ignorance of cultivating leading industries and weak employment ability. Statistics show that in the existing 19,000 small towns, the average employment for the urban population of the total population is 26%, and employment accounting for 56% of the total pupulation in cities with a population

of 200,000–500,000. We should pay attention to culture, especially in industry and agricultural-related industries to support the construction of cities and towns. It is necessary to attach great importance to the role of private economy in urbanization.

2. *The principle of agglomeration effects:* The development of cities and towns should be "industry popular" and "human popular." Industrial layout should be concentrated; urban development should focus on population concentration, all of which are good to the flow of goods. We should set up mechanisms for flow of talent to encourage different kinds of people participating in the construction of rural towns to solve the problem of insufficient personnel during the urbanization process. In order to guide enterprises, capital, and labor population to concentrate on the cities and towns, the government has to comply with the development of enterprises and the requirements of cities and towns and do a good job in planning, organization, coordination and services, and create good business environment for enterprises by sharing infrastructure which enables enterprises to take advantage of being immune to external forces; enable enterprises to reduce transaction costs through concentration and operating costs.
3. *The principle of function:* Cities and towns should strengthen water and electricity supply, transport, culture, and entertainment facilities and focus on the urban infrastructure of hardware and software environment, be concerned about the harmonious development of urban economy, society, ecology, culture, environment, and education. We should adhere to the concept of sustainable development of ecological environment and natural resources protection in the process of urbanization.

12.5.4.3 Vigorous Promotion of the Public Life of Farmers and Achieving the Goal of Production, Life, and Eco-Efficiency, Promotion of the Process of China's Industrial Structure Upgrading With further deepening of China's reforms, there will be increasing rural urbanization. The development of small towns can transfer a large number of surplus rural labor force to the urban centers. There are two ways of changing China's population of farmers to nonagricultural population; the first is to transfer the transforming population of farmers to large and medium-sized cities. They are mainly college graduates with higher education and businessmen. The second is to absorb surplus rural labor force through the construction of cities and towns. This is the most effective way for the urbanization of rural people. "Leave earth but not home" is the main direction of transferring surplus rural labor force, which can avoid the problem of excessive number of emerging prosperous cities and long-term slump of rural areas.

Effectively realizing urbanization of the rural population, we should solve the problem of employment first. A large number of rural population transferring to nonagricultural industries and urban areas, should be assured of sufficient employment opportunities. Through internal industrial structure adjustment of agriculture and fully open urban labor market, we should fully display the potential of various economic sectors, especially the private economy sectors in the area of employment, which show huge potential of the urban tertiary industries to absorb agricultural labor. Second, we should correctly handle the issue of urban construction. The migration of rural population to urban regions will surely raise new requirements of infrastructure in cities and towns, especially the urban housing system. The state, collectives, and also individuals should be charged with the construction of cities and towns. The third is the issue of household registration system. We should ensure that rural residents and urban residents enjoy similar subsidy system for basic old-age insurance, basic medical insurance system, and minimum living security systems. The fourth problem is the issue of job training for rural residents. We should establish surplus rural labor force training and employment guidance mechanisms to strengthen job training of rural residents in order to overcome difficulties in finding jobs and the smooth job transfer to the secondary and tertiary industries. So the majority of rural residents will enjoy the benefits of urbanization and the colorful urban culture.

Bibliography

Akamatsu. 1961. A theory of unbalanced growth in the world economy. In: *Weltwirtschaftliches* Archiv, Hamburg, no. 86, pp. 196–217.

Akamatsu, Kaname. 1962. A historical pattern of economic growth in developing countries. *Journal of Developing Economies* 1: 3–25.

Applegate, Heidi F. 1996. Technology or economics system: an analysis of industrial structure in the G.D.R. The Pennsylvania State University.

Arndt, Sven W. and Kierzkowski Henryk. 2001. *Fragmentation: New Production Patterns in the World Economy*. New York: Oxford University Press.

Carlton, Dennis W. and Jeffrey M. Perloff. 1990. *Modern Industrial Organization*. Glenview, IL: Scott Foresman.

Chenery, Hollis Burnley 1979. *Structural Change and Development Policy*. New York: Oxford University Press.

Chenery, Hollis Burnley and Moises Syrquin. 1975. *Patterns of Development*. New York: Oxford University Press.

Chenery, Hollis Burnley, Sherman Robinson, and Moises Syrquin. 1986. *Industrialization and Growth: A Comparative Study*. New York, NY: Oxford University Press.

Choi, Younki. 2000. Industrial structure, growth and income disparity. Buffalo: State University of New York at Buffalo.

Chu, Yanfeng and Sifeng Liu. 2007. Input-output analysis of logistic industries in China. *Enterprise Economy* 12: 70–72.

Dang, Yaoguo and Sifeng Liu. 1995. The shift and disposition of surplus labourers in rural areas in Henan Province. *Acta Agriculturae Universitatis Henanensis* 29(3): 223–227.

Dang, Yaoguo and Sifeng Liu. 2004. Study on grey synthetic clusters appraisals model, *IEEE SMC*: 2398–2402.

Dang, Yaoguo, Qingye Zhao, and Sifeng Liu. 2000. Selection and construction of evaluation index system for agricultural leading industries. *Journal of Agricultural Economics* 1: 6–9.

Dang, Yaoguo, Sifeng Liu, Bin Liu, et al. 2005. Study on grey incidence decision model of the dynamic multiple-attribute. *Engineering Science* 7(2): 69–72.

Dang, Yaoguo, Sifeng Liu, Bin Liu, et al. 2005. Study on the integrated grey clustering method under the clustering coefficient with non-distinguished difference. *Chinese Journal of Management Science* 13(4): 69–73.

Dang, Yaoguo, Sifeng Liu, Bingjun Li, et al. 2000. Measurement and calculation on the contribution rate of the technological advance of the primary industry in Henan Province. *Systems Engineering—Theory and Practice* 8: 95–99.

Dang, Yaoguo, Sifeng Liu, Kejia Chen, et al. 2005. Structure adjustment of the tertiary industry in Jiangsu Province and the "turnpike way" Model. *Statistics and Decision* 4: 96–98.

Dang, Yaoguo, Sifeng Liu, Xinli Xue, et al. 2006. Mathematical models of the adjustment of the internal structure for the Chinese primary industry and responding measure. *Systems Science and Comprehensive Studies in Agriculture* 22(2): 81–83.

Dang, Yaoguo, Sifeng Liu, Yaojun Ye, et al. 2000. The theory and practice in the management of the industrialization of agriculture in Henan Province. *System Sciences and Comprehensive Studies in Agriculture* 16(3): 187–190.

Dang, Yaoguo, Sifeng Liu, Ying Wang, et al. 2002. Mathematical model and strategy analysis of adjustment of the internal structure of the primary industry in Jiangsu Province. *System Sciences and Comprehensive Studies in Agriculture* 18(4): 241–248.

Dang, Yaoguo, Sifeng Liu, Zhuo Zhang, et al. 2003. Structure adjustment of the secondary industry in Jiangsu Province and the "turnpike way" model. *Journal of Southern Yangtze University* (Natural Science Edition) 2(5): 526–530.

Dang, Yaoguo, Sifeng Liu, and Jianping Wang. 2000. Study on patterns of development for agriculture's industrial operation in Henan Province. *Journal of Agricultural Economics* 5: 48–50.

Dang, Yaoguo, Sifeng Liu, and Kejia Chen. 2003. Fast track model of industrial structure adjustment and its Empirical analysis. *Modern Economic Research* 5: 10–12.

Dang, Yaoguo, Sifeng Liu, Zhenjie Zhai, et al. 2004. Study on model of grey comprehensive cluster. *Statistics and Decision* 10: 4–5.

Dang, Yaoguo, Sifeng Liu, and Zhenjie Zhai. 2004. How to choose the regional leading industries. *Economic Survey* 6: 38–40.

Fang, Jia. 1997. *Study on Industrial Structure Problems*. Beijing, China: China Renmin University Press.

Fang, Zhigeng, Sifeng Liu, Han Zhang, et al. 2004. Analysis and trend prediction of main economy development indexes of Nanjing city during "the 11th five-year plan". *Social Sciences in Nanjing* 1: 31–40.

Feng, Lixiang, Sifeng Liu, Hongxing Shi, et al. 2008. Selection of regional leading industries based on maximum entropy law. *Industrial Technology and Economy* 8: 64–67.

Fu, Qiang and Xiutai Yang. 2000. The dynamic model describing influence of industrial structure and technology transfer on economy growth. *Journal of Chongqing University* (Natural Science Edition) 3: 43–46.

Gabriele, A. 2001. Science and technology policies, industrial reform and technical progress in China. In: *United Nations Conference on Trade and Development*. Discussion papers, no. 155.

Gan, Zhihe. 2000. *Study on Development and Adjustment of Industrial Structure*. Beijing, China: China Economics Press House.

Gong, Yangjun. 2002. *Study on Industrial Structure*. Shanghai, China: Shanghai University of Finance and Economics Press.

Guo, Kesha and Yanzhong Wang. 1999. *Study on Policy and Changing Trend of China Industrial Structure*. Beijing, China: Economy and Management Publishing House.

Guo, Kesha. 1995. Study on changing trend and policy of China industrial structure. *Management World* 5: 73–83.

Henderson, W.O. and Chaloner, W.H. (translated). 1958. *The Growth of Industrial Economies*. By W.G. Hoffman (Ed.), Manchester: Manchester University Press, pp. 166–170.

Hong, Yinxing. 2001. China's trade structure adjustment and industrial escalation under WTO Conditions. *Management World* 2: 21–26.

Hu, Rongtao. 2001. *Analysis of Industrial Structure and Regional Economic Benefit*. Beijing, China: Economy and Management Publishing House.

Jiang, Xiaojuan. 2005. The improvement and upgrade of industrial structure: new tasks in the new stage. *Finance and Trade Economics* 4: 3–9.

Jiang, Zehua and Yan Bai. Analysis of influence factors and connotation for industrial structure updating. *Contemporary Economic Research* 10: 53–56.

Jiang, Zhaoxia. 2005. *Study on Industrial Structure Problems*. Beijing, China: China Economics Press House.

Ko, Posheng. 1996. Dynamics of industrial structure in a small semi-open economy. The University of Utah.

Kuznets, Simon Smith. 1989. *Modern Economic Growth*. Beijing, China: Beijing Economy Academy Press.

Kuznets, Simon Smith. 1999. *Economic Growth of Nations: Total Output and Production Structure*. Beijing, China: The Commercial Press.

Leontief, Wassily. 1986. *Input-Output Economics*, 2nd edition. New York, NY: Oxford University Press.

Li, Bingjun and Junjuan Liu. 2004. Comparison and analysis of the action effect of capital element to the economic development in the different regions. *East China Economic Management* 18(2): 30–32.

Li, Bingjun and Wenlong Zhang. 2006. Analysis on regional development variance and growth convergence of Henan Province. *East China Economic Management* 20(2): 39–42.

Li, Bingjun, Xiaoyan Qiao, and Sifeng Liu. 1999. Quantitative method to determine agricultural leading industries and its application. *Acta Agriculturae Universitatis Henansis* 33: 43–46.

Li, Jinghua. 2001. The Markov chains for industrial structure transition in China. *Mathematics in Practice and Theory* 2: 156–161.

Li, Qiaoxing, Sifeng Liu, and Yaoguo Dang. 2007. Extension analysis to adjust the industrial structure of rural economy in China. *Journal of Harbin Institute of Technology* 7: 1205–1208.

Li, Qiaoxing, Sifeng Liu, and Yi Tan. 2007. The positive character vector method to adjust grey direct consumption coefficients using the dependent function. *Journal of Guangxi University* (Natural Science Edition) 4: 346–349.

Li, Qiaoxing, Sifeng Liu, and Ying Li. 2006. Study of existence of solution to static non-linear input-output model with polynomial function. *Operations Research and Management Science* 2: 91–93.

Li, Sha and Sifeng Liu. 2006. Impact on competition for IT industry clustering in Jiangsu Province. *Statistics and Decision* 19: 96–98.

Li, Sha and Sifeng Liu. 2007. Cost analysis of high and new technological industries in China. *Statistics and Decision* 4: 64–66.

Li, Xiaoting. 2002. On strategic choice in our country's industrial structure optimization. *Journal of China Agricultural University* (Social Science Edition) 1: 52–56.

Lin, Chi-Rung Chelsea. 2001. Essays on political economy of international trade and industrial structure. University of California.

Liu, Bin, Rong Zhang, Zhenjie Zhai, et al. 2005. The optimal GM(1,1) and its application in forecasting the diversion amount of rural labors. *System Sciences and Comprehensive Studies in Agriculture* 3: 175–177.

Liu, Jian. 2000. *Updating of China Industrial Structure.* Beijing, China: China Yanshi Publishing House.

Liu, Junjuan and Bingjun Li. 2005. Adjustment of industrial structure based on fast track model in Henan Province. *Industrial Technology and Economy* 6: 59–62.

Liu, Sifeng and Bingjun Li. 1998. Study on the mathematical model and indices for the synthetic evaluation of regional leading industry. *Chinese Journal of Management Science* 6(2): 8–13.

Liu, Sifeng and Julong Deng. 2000. The range suitable for GM(1,1). *Systems Engineering—Theory and Practice* 5: 121–124.

Liu, Sifeng and Ruilan Wang. 2003. Grey incidence analysis of the third industry in the "ninth five-year plan" period in Nanjing. *Journal of Nanjing University of Science and Technology* 16(5): 55–58.

Liu, Sifeng and Yaoguo Dang. 2004. Technical change and the funds for science and technical advance. *The International Journal of Systems & Cybernetics* 2: 295–302.

Liu, Sifeng and Yaoguo Dang. 2005. *Prediction Method and Technology*. Beijing, China: High Education Press.

Liu, Sifeng and Yi Lin. 1998. *An Introduction to Grey Systems: Foundations, Methodology and Applications*. Slippery Rock: IIGSS Academic Publisher.

Liu, Sifeng and Yongda Zhu. 1993. Study on triangular model and indexes in synthetic evaluation of regional economy. *Transactions of the Chinese Society of Agricultural Engineering* 9(2): 8–13.

Liu, Sifeng, Bingjun Li, Xiuli Li, et al. 1999. Contribution rate of technological advance of Henan Province and the lateral comparison. *Acta Agriculturae Universitatis Henanensis* 33(1): 40–43.

Liu, Sifeng, Xuewen Tang, Chaoqing Yuan, et al. 2004. Study on degree of order of China industrial structure. *Economic Perspectives* 5: 53–56.

Liu, Sifeng, Yaoguo Dang and Zhigeng Fang. 2004. *Grey Systems Theory and Its Applications*, 3rd edition. Beijing, China: Science Press.

Liu, Sifeng, Yaoguo Dang, Bingjun Li, et al. 1999. The G-C-D model and measurement of the contribution rate of technological advance. *Chinese Journal of Management Science* 7(2): 76–80.

Liu, Sifeng, Yaoguo Dang, and Bingjun Li. 1999. A positive study on the relationship between technological advancement and input for science and technology in Henan Province. *Acta Agriculturae Universitatis Henanensis* 33(4): 339–341.

Liu, Sifeng. 1991. Study on optimization of the rural economic structure of Henan Province. *Journal of Henan Agriculture University* 25(1): 38–44.

Liu, Zhibiao, Guosheng Wang, and Tongliang An. 2001. *Analysis of Modern Industrial Economy*. Nanjing, China: Nanjing University Press.

Luo, Dang and Sifeng Liu. 2004. On the algorithm of grey multiple objective programming. *Systems Engineering* 5: 12–15.

Luo, Dang and Sifeng Liu. 2004. Study on grey dynamic programming. *Systems Engineering—Theory and Practice* 4: 56–62.

Luo, Dang and Yuhui Qin. 2007. Grey clustering method based on attribute recognition. *Journal of North China Institute of Water Conservancy and Hydroelectric Power* 5: 90–92.

Luo, Dang, Sifeng Liu, and Yaoguo Dang. 2003. The optimization of grey model GM(1,1). *Engineering Science* 8: 50–53.

Luo, Dang. 2007. The method for grey interval clustering decision. *Journal of Zhengzhou University* (Natural Science Edition) 1: 119–124.

Mao, Jian. 1999. *Industrial Structure Changing and Selection of Industrial Policy*. Beijing, China: China Financial and Economic Publishing House.

Mclaren, John. 2000. "Globalization" and vertical structure. *American Economic Review* 5: 1239–1254.

Mi, Chuanmin, Sifeng Liu, and Ju Yang. 2004. Study on grey incidence between scientific input and economic growth in Jiangsu Province. *Science of Science and Management of Science and Technology* 1: 34–36.

Mi, Chuanmin, Sifeng Liu, Yaoguo Dang, et al. 2006. Study on grey entropy weight clustering decision-making. *Systems Engineering and Electronics* 12: 1823–1825.

Ning, Xuanxi and Sifeng Liu. 2003. *Prediction Method and Technology*. Beijing, China: Science Press.

Ren, Junqing and Sifeng Liu. 2008. Analysis on supply chain stability based on grey model of input and output. *Journal of Nanjing University of Aeronautics and Astronautics* (Social Science Edition) 1: 21–24.

Ren, Yingying and Sifeng Liu. 2006. Grey incidence analysis about influencing factors on industrial structure evolution in Jiangsu Province. *Commercial Research* 14: 48–51.

Ren, Yingying, Sifeng Liu, and Zhigeng Fang. 2006. Construction of conception system for scientific technology production forces. *Science and Technology Progress and Policy* 8: 120–122.

Rostow, W.W. 1960. *The Stages of Economic Growth: A Non-Communist Manifesto*. Cambridge: Cambridge University Press, pp. 4–16.

Ruan, Aiqing and Sifeng Liu. 2008. Study on the life cycle of industry cluster by evolutionary game chain model. *Science of Science and Management of Science and Technology* 2: 91–95.

Ruhashyankiko, Jean-Francois. 1999. Essays on international changes in industrial structure of production and trade. Boston: Harvard University.

Ryutaro, Komiya. 1988. *Industrial Policy of Japan*. Beijing, China: Academic Press.

Santiago, Luis Enrique. 1999. Industrial structure as a determinant of economic growth. Cornell University.

Shi, Hongxing, Sifeng Liu, and Zhigeng Fang. 2007. Research on flowing mechanism of the technical productive forces. *Science of Science and Management of Science and Technology* 11: 25–28.

Sun, Fuquan. 2006. *Microadjustment of Industrial Structure*. Beijing, China: China Economics Press House.

Sun, Jie and Jian Yu. 2007. Adjustment of China industrial structure under condition of opening economics. Beijing, China: Economy and Management Publishing House.

Tang, Xuewen and Xiumei Lu. 2003. Adjustment and development of the tertiary industry in Jiangsu Province. *Modern Economic Research* 11: 67–70.

Tirole, Jean. 1988. *The Theory of Industrial Organization*. Boston, MA: The MIT Press.

Wang, Qingfeng. 2006. Do well in the brand construction to strengthen rural labor economy. *Career Horizon* 6: 3–5.

Wang, Qingfeng. 2007. Study on path of regional brand training of food industry in Henan Province. *Enterprise Vitality* 2: 27–29.

Wang, Qingfeng. 2007. Study on service brand construction of logistic enterprise. *China Market* 15: 4–5.

Wang, Qingfeng. 2008. Development of regional industrial cluster and leading enterprise. *Enterprise Vitality* 3: 76–77.

Wang, Ruilan and Sifeng Liu. 2005. Analysis on advantage and disadvantage of industry's identical structure. *Statistics and Decision* 12: 124–125.

Wang, Ruilan, Sifeng Liu, and Yaoguo Dang. 2005. Study on indexes about constructing regional scientific development target. *Forum on Science and Technology in China* 6: 79–82.

Wang, Ying and Sifeng Liu. 2003. Grey incidence analysis between degree of open and China economic growth. *Practice in Foreign Economic Relations and Trade* 2: 5–7.

Wang, Ying and Sifeng Liu. 2008. An empirical analysis on the channels of international technology spillovers. *The Journal of Quantitative and Technical Economics* 4: 153–161.

Wang, Ying and Sifeng Liu. 2008. The influence of OFDI on Chinese industrial structure: an analysis based on grey incidence theory. *World Economy Study* 4: 61–65.

Wang, Yueping. *Updating of Industrial Structure under the Condition of Opening Economy.* Beijing, China: Economy and Management Publishing House.

Wong, John and Sarah Chan. 2002. China's emergence as a global manufacturing centre: implications for ASEAN. *Asia Pacific Business Review* 1: 79–94.

Wu, Hecheng and Chuiyong Zheng. 2004. Exponential distribution-based input-output model. *Journal of Hehai University* (Natural Science Edition) 2: 233–235.

Wu, Hecheng and Chuiyong Zheng. 2005. Study on stochastic input-output models. *Journal of Hehai University* (Natural Science Edition) 3: 347–349.

Wu, Hecheng and Sifeng Liu. 2007. Evaluation on the R&D relative efficiency of different areas in China based on improved DEA model. *R&D Management* 2: 108–112.

Wu, Jinhong. 2007. *Opening Economics and Industrial Structure Updating.* Beijing, China: Social Science Academic Press.

Xie, Naiming and Sifeng Liu. 2006. Research on discrete grey model and its predictive result. *Journal of Systems Engineering* 5: 520–523.

Xue, Jingxiao and Tianbao Zhang. 2002. Characteristic and general style of technology progress promoting industrial structure updating. *Forum of World Economy and Politics* 4: 10–14.

Xue, Xinli, Xuewen Tang, and Yaoguo Dang. 2005. Elasticity analysis between employment and development of the tertiary industry in Jiangsu Province. *Statistics and Decision* 4: 127–128.

Yang, Jianmei. 2003. Sky-ine diagram—a synthesis approach to industrial structure study and its application. *Systems Engineering—Theory and Practice* 12: 10–17.

Yang, Wenhan and Sifeng Liu. 2006. Relationship study of technology innovation and interaction between SC enterprises. *Studies in Science of Science* 5: 798–802.

Yin, Jianhua. 2000. Product structure adjustment and industry structure optimization on economy increase theory. *Systems Engineering—Theory and Methodology Applications* 2: 168–172.

Ying, Longgen. 1999. China's changing regional disparities during the reform period. *Economic Geography* 1: 59–70.

Zhang, Min, Yaoguo Dang, and Rui Wang. 2007. Policy suggestion and prediction analysis on demand for scientific and technological talents during "the 11th five-year plan" in Jiangsu Province. *Science and Technology Progress and Policy* 5: 165–167.

Zhang, Shixian. 2000. Empirical study on industrial investment efficiency and industrial structure changing. *Management World* 5: 79–85.

Zhou, Chunguang and Sifeng Liu. 2006. Analysis of contribution effect of foreign direct investment to product export. *Management World* 1: 145–146.

Zhou, Zhenhua. 1995. *Structural Effects in Modern Economic Growth*. Shanghai, China: Shanghai People's Publishing House.

Zhu, Jianjun, Weili Wu, and Sifeng Liu. 2006. Subjective-objective combination estimation weight based on fuzzy linear programming. *Operations Research and Management Science* 3: 19–24.

Zhu, Qiang. 2002. Economic globalization and evolution of leading industries in developing countries. *Forum of World Economy and Politics* 6: 7–11.

Zhu, Yongda and Sifeng Liu. 1997. *Theory, Model and Applications of Middle-Economic Management*. Wuhan, China: Huazhong University of Science and Technology Press.

Index

B

C

K

L

Q

R

S

£93.00